Ökonometrie und Unternehmensforschung

Econometrics and Operations Research

IX

Herausgegeben von / Edited by

M. Beckmann, Bonn · R. Henn, Karlsruhe · A. Jaeger, Cincinnati
W. Krelle, Bonn · H. P. Künzi, Zürich
K. Wenke, Ludwigshafen · Ph. Wolfe, Santa Monica

Geschäftsführende Herausgeber / Managing Editors
W. Krelle · H. P. Künzi

Martin J. Beckmann

Dynamic Programming of Economic Decisions

Springer-Verlag New York Inc. 1968

Professor Dr. MARTIN J. BECKMANN

Institut für Ökonometrie und Unternehmensforschung der Universität Bonn
and Brown University, Providence, R. I.

ISBN 978-3-642-86451-3   ISBN 978-3-642-86449-0 (eBook)
DOI 10.1007/978-3-642-86449-0

Title-No. 6484

Parentibus et
Soceris

Preface

Dynamic Programming is the analysis of multistage decision in the sequential mode. It is now widely recognized as a tool of great versatility and power, and is applied to an increasing extent in all phases of economic analysis, operations research, technology, and also in mathematical theory itself. In economics and operations research its impact may someday rival that of linear programming.

The importance of this field is made apparent through a growing number of publications. Foremost among these is the pioneering work of Bellman. It was he who originated the basic ideas, formulated the principle of optimality, recognized its power, coined the terminology, and developed many of the present applications. Since then mathematicians, statisticians, operations researchers, and economists have come in, laying more rigorous foundations [KARLIN, BLACKWELL], and developing in depth such application as to the control of stochastic processes [HOWARD, JEWELL]. The field of inventory control has almost split off as an independent branch of Dynamic Programming on which a great deal of effort has been expended [ARROW, KARLIN, SCARF], [WHITIN], [WAGNER]. Dynamic Programming is also playing an increasing role in modern mathematical control theory [BELLMAN, Adaptive Control Processes (1961)]. Some of the most exciting work is going on in adaptive programming which is closely related to sequential statistical analysis, particularly in its Bayesian form.

In this monograph the reader is introduced to the basic ideas of Dynamic Programming. An attempt is made to show the breadth of this method by a careful exposition of its principles followed by some typical applications to economic analysis, operations research and decisions in general. Some effort has been made to give a systematic exposition of the various possibilities: discrete and continuous sequences; discrete and continuous decision variables; certainty, risk, and uncertainty. Needless to say, this book cannot cover all aspects of Dynamic Programming— the expansion of the field is much too rapid for that. Such interesting areas as combinatorial problems, stopping rules, statistical decision theory, and control theory are not included. Emphasis is on those parts which appear to be most immediately applicable to economic analysis.

The writing up of this book was made possible by a sabbatical leave from Brown University for which I should like to express my deep appreciation. I should also like to thank the Deutsche Forschungsgemeinschaft for two travel grants in support of required research and RICHARD BELLMAN for some interesting and stimulating conversations. Some of the material presented here was first developed for a series of invited lectures at the Faculté Polytechnique de Mons. I should like to thank my Belgian colleagues for this invitation and their stimulating comments. Finally I want to thank my assistant Dr. HOCHSTÄDTER for his help in calculating various examples and in reading and correcting the manuscript, Dr. KASPAR for locating references, GOETZ UEBE and RALF MOOCK for proofreading, and Miss HERBST for her diligent typing which made the rapid completion of this book possible.

Bonn, May 1968   MARTIN J. BECKMANN

General Literatur

OR	Operations Research
NRLQ	Naval Research Logistics Quarterly
Ann. Math. Stat.	Annals of Mathematical Statistics
MS	Management Science
J. Ind. Eng.	Journal of Industrial Engineering
JMM	Journal of Math. Mechanics
ORQ	Operations Research Quarterly
JPE	Journal of Political Economy
SIAM	Society for Industrial and Applied Mathematics
Journal Math. Anal. Appl.	Journal of Mathematical Analysis and Aplications
Proc. Nat. Acad. Sciences	Proceedings of the National Acadamy of Sciences
Bull. Soc. Royale Sciences Liège	Bulletin de la Societé Royale des Sciences de Liège
ch.	chapter
p.	page

Books

ARIS, RUTHERFORD: Discrete Dynamic Programming. An Introduction to the Optimization of Staged Processes. New York—Toronto—London: Blaisdell 1964.

BELLMAN, R. E.: Adaptive Control Processes: A Guided Tour. Princeton: Princeton University Press 1961.

— Dynamic Programming. Princeton: Princeton University Press 1957.

—, and E. DREYFUS: Applied Dynamic Programming. Princeton: Princeton University Press 1962.

—, and R. KALABA: Dynamic Programming and Modern Control Theory. New York—London: Academic Press 1965.

—, and R. KARUSH: Dynamic Programming: A Bibliography of Theory and Application. Santa Monica: RAND Corp. 1964. Memorandum RM 3951 PR, Febr. 1964. Memorandum RM 3951 1 PR, Aug. 1964.

DANTZIG, G. B.: Linear Programming and Extensions (Ch. 23). Princeton: Princeton University Press 1963.

DREYFUS, S. E.: Dynamic Programming and the Calculus of Variations. New York: Academic Press 1965.

ELSNER, K.: Mehrstufige Produktionstheorie und dynamisches Programmieren. Meisenheim am Glan: Verlag Anton Hain. 1964.

GIBSON, J. E.: Proceedings of Dynamic Programming Workshop. Lafayette, Indiana: Purdue University 1961.

HADLEY, G.: Nonlinear and Dynamic Programming. London: Addison-Wesley 1964.

KAUFMANN, A., and R. CRUON: Dynamic Programming—Sequential Scientific Management. New York—London: Academic Press 1967.

NEMHAUSER, G. L.: Introduction to Dynamic Programming. New York: Wiley 1966.
NOTON, A. R. M.: Introduction to Variational Methods in Control Engineering. Oxford—London—New York: Pergamon Press 1965.
PIEHLER, Joachim: Einführung in die Dynamische Optimierung (Dynamische Programmierung). Leipzig: Teubner 1966.
SAMUELSON, P. A.: Economics. 7th ed. New York: McGraw Hill 1967.
WENTZEL, J. S.: Elemente der Dynamischen Programmierung. München: Oldenbourg Verlag 1966.

Papers

BELLMAN, R. E.: Dynamic Programming and a New Formalism in the Calculus of Variations. Proc. Nat. Acad. Sciences, USA, **40,** 4, 231—235 (1954).
— The Theory of Dynamic Programming. In: Modern Mathematics for the Engineer, ed. E. F. Beckenbach, Ch. 11, p. 243—278. New York: McGraw Hill Book Comp. 1956.
DREYFUS, S.: Dynamic Programming. In: Progress in Operations Research, ed. by Russel L. Ackoff. Vol. I, 211—242. New York: Wiley 1961.
HOWARD, R. A.: Dynamic Programming. MS **12,** 5, 317—348 (1966).
KIMBALL, G. E., and R. A. HOWARD: Sequential Decision Processes. In: Notes on Operations Research 1959, assembled by the O. R. Center, MIT. Cambridge, Mass.: The Technology Press 1959.
SASIENI, M., A. YASPAN, and L. FRIEDMAN: Operations Research: Methods and Problems. Ch. I, p. 270—293. New York: Wiley 1959.
SIMPSON, M. G., C. S. SMITH, H. G. BASS, and D. J. KIRBY: Dynamic Programming. In: Some Techniques of Operational Research, ed. by B. I. Houlden. London: The English Universities Press 1962.

Contents

Preface VII

General Literature IX

Part One.   Finite Alternatives

§ 1. Introduction 1
§ 2. Geometric Interpretation 6
§ 3. Principle of Optimality 10
§ 4. Value Functions for Infinite Horizons: Value Iteration 15
§ 5. Policy Iteration 19
§ 6. Stability Properties 21
§ 7. Problems without Discount and with Infinite Horizon 24
§ 8. Automobile Replacement 26
§ 9. Linear Programming and Dynamic Programming 33
        References and Selected Reading to Part One 35

Part Two.   Risk

§ 10. Basic Concepts 37
§ 11. The Value Function 41
§ 12. The Principle of Optimality 45
§ 13. Policy Iteration 48
§ 14. Stability Properties 51
§ 15. Solution by Linear Programming 54
§ 16. Machine Care 55
§ 17. Inventory Control 61
§ 18. Uncertainty: Adaptive Programming 76
§ 19. Exponential Weighting 81
        References and Selected Reading to Part Two 83

Part Three.   Continuous Decision Variable

§ 20. An Allocation Problem 85
§ 21. General Theory 87
§ 22. Linear Inhomogeneous Problems 94
§ 23. A Turnpike Theorem 96
§ 24. Sequential Programming 99
§ 25. Risk 100
§ 26. Quadratic Criterion Function 101
        References and Selected Reading to Part Three 104

Part Four.   Decision Processes in Continuous Time

§ 27. Discrete Action 107
§ 28. Variable Level 110
§ 29. Risk of Termination 112

§ 30.   Discontinuous Processes—Repetitive Decisions   113
§ 31.   Continuous Time Inventory Control   120
§ 32.   Continuous Action—Steady State Problems   123
§ 33.   The Principle of Optimality in Differential Equation Form   125
§ 34.   Dynamic Programming and the Calculus of Variations   129
§ 35.   Variation under Constraints: The Maximum Principle   132
        References and Selected Reading to Part Four   135

Author Index   136

Subject Index   138

PART ONE

Finite Alternatives

§ 1. Introduction

Dynamic Programming (DP) is the analysis of maximum problems in recursive form. Mathematically it is the study of functional equations which contain a maximum (or minimum or min max) operator. The motivation for this technique may be seen by a closer consideration of decision processes.

Decision Processes. Following ABRAHAM WALD [WALD] a decision may be represented by the following schema.

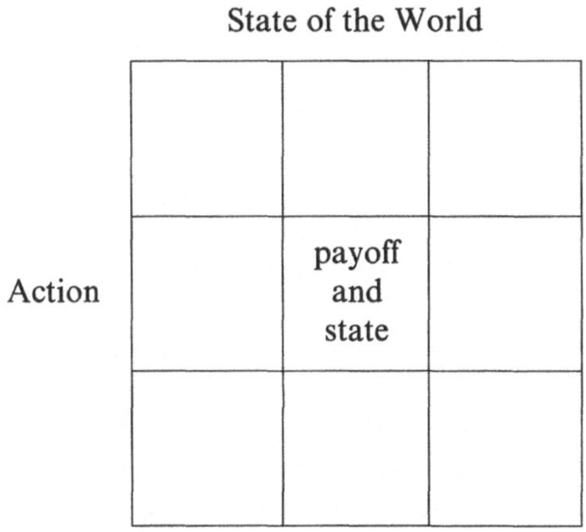

State of the World

	payoff and state	

Fig. 1

A decision is an action in a state of the world. It generates an outcome which is evaluated in terms of a payoff. While the schema in this generality may be used to represent complex sequences as well, it is often convenient for purposes of analysis to break a decision problem down into simpler steps to be analysed in sequence. This is achieved by assigning to an action in a state of the world an outcome composed not only of a payoff but also of a new state of the world. The new state, which may be identical with the previous one, gives rise to a new decision.

A finite sequence of decision can arise only when there are states in which no further decision is required (terminating states) or when the number of decisions to be made in sequence is limited in advance (horizon). In principle the chain of decisions may go on forever. Termination may also be a random event e. g. with a constant probability p the decision chain is terminated after each step. (The length of the decision sequence is then subject to a geometric distribution.)

The arrangement of decisions need not be temporal but may be spatial—a network (see below)—or purely logical. This means that the breaking down of a decision into steps can be a problem in itself.

Example: The KNAPSACK Problem.

Suppose the following items are useful on a hike but that the hiker does not want to carry more than twenty kilograms and not more than one of each item.

Table 1

items	weight g_i	utility a_i
1. tent	10	10
2. sleeping bag	5	30
3. canned goods	10	15
4. bread	1	15
5. hatchet	4	20
6. salami	2	10

In this problem stages may be defined, for instance, by considering first only solutions involving one commodity, next those with at most two commodities etc. After the one-step problem has been solved then the maximum utility is known that can be achieved by carrying one item weighting no more than x kg, say $v_1(x)$. With the aid of this function, the two-item problem can be reduced to a one-item problem by determining for each weight x

$$\text{Max}_i \left[a_i + v_1(x - g_i) \right].$$

Here a_i denotes the utility, g_i the weight of commodity i. This maximization is subject to the restriction that i is not contained among the goods, which generated $v_1(x - g_i)$. In this way one obtains a new function $v_2(x)$ showing the maximum utility achievable with two items under a total weight limit of x. Continuing in this way one eventually determines $v_6(x)$ and solves the original problem from $v_6(20)$.

If it turns out that $v_5(x) = v_4(x)$ for all $x \leq 20$, then one should never want to take more than four items along and the problem has been solved already at stage 4.

In principle feedback loops and more complicated concatenations of decisions are also possible. But in this book we shall restrict ourselves to simple linear sequences of decisions [cf. NEMHAUSER (1966)].

Formally a sequential decision problem in linear sequence may be described as follows: at the first stage several alternative actions are possible which at the second stage give rise to further alternatives etc. until in any problem with a finite number of possibilities each branch is terminated. Graphically this represents a tree (Fig. 2). Each node corresponds to a stage, i.e. a possible state of the problem. Each branch represents a transition, i.e. a decision. Associated with each transition and possibly also with the states themselves are costs and returns or — as we shall say from now on — utilities (returns minus costs when measured in money). For each sequence of decisions there is defined a total utility, i.e. the sum of utilities of the individual steps. This assumption of additivity is crucial for the applicability of the DP technique. This total utility may be assigned to the terminal points of the various branches of the decision tree as the unique representatives of the various possible decision sequences.

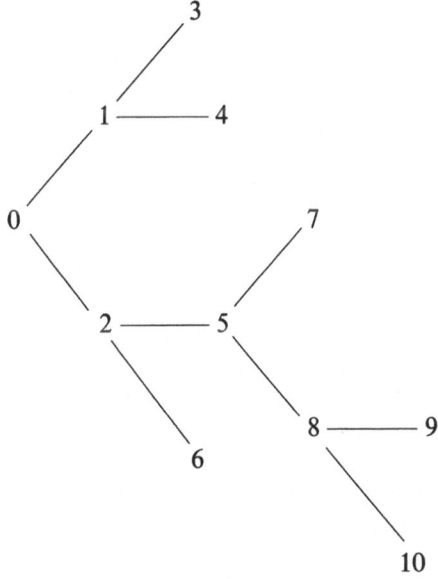

Fig. 2

1*

Optimal sequences. We have just shown that each state gives rise to a finite number of possible decision chains of finite length, and that with each chain of decisions a sum of payoffs is associated. A decision sequence will be called optimal if it has maximum total utility, i. e., if it leads to an endpoint to which is attached the highest total utility. It is realized by the sequence of decision steps leading to this terminal point.

In a large tree the comparison of total utilities may be tedious. It can be done in a systematic way by moving first forward and adding up the utility sums starting from the initial state at zero, say, along the various routes and noting the intermediate sums at the various nodes. In this way the totals for the various endpoints may be determined with a minimum of duplication. Next going backward from these terminal points at each branch point the highest total utility may be registered which has been generated by any sequence leading on from there. In this way maximum utilities are associated with the *starting* points of decision chains (rather than termination points, as before). Moving backwards from node to node one eventually finds the highest utility obtainable from node zero.

It turns out that the forward calculation may be omitted since it is possible to carry out the addition and comparison of utilities in one movement. One begins by setting utilities in terminal states zero or possibly equal to given terminal payoffs. For the states which are one step removed from these terminal states, utilities of the various transitions plus terminal utilities are calculated and compared. For every node the largest utility sum achievable in any direction is noted down. This is the value function or utility assigned to that node. This operation is repeated following the schema

(1) $$v_i = \underset{j \in S_i}{\text{Max}} \left[a_{ij} + v_j \right].$$

Here S_i is the set of next nodes reached from state i, a_{ij} the payoff of the transition $i \rightarrow j$, and v_j the maximum utility realized by chains starting in state j. Moving backward in this way the largest utility sum that can be reached is determined for every i and eventually for the initial state $i = 0$. These utilities may be called the values of states. In this way a value function v_i is defined over the set of states. We have described in a nutshell the technique of recursive maximization as applied to decision trees.

These intuitive ideas may be described formally as follows: A (finite) chain of decisions is called optimal if it achieves the largest utility sum. This sum of utilities called the value of state i, v_i

$$v_i = \underset{\substack{\text{all chains} \\ \text{from } i}}{\text{Max}} \left(\sum_{\text{chains}} a_{jk} \right).$$

Note that even when the optimal decision chains are not unique the value of a state is uniquely determined.

Denote the states in an optimal sequence starting at i by $j_1, j_2, \ldots, j_n$. Then

$$v_i = a_{ij_1} + a_{j_1 j_2} + \cdots + a_{j_{n-1} j_n} + a_{j_n}$$

$$(2) \qquad \qquad = a_{ij_1} + v_{j_1}$$

$$\geq a_{ik_1} + a_{k_1 k_2} + \cdots + a_{k_{m-1} k_m} + a_{k_m}$$

for any other decision chain from i and in particular for those $k_1, k_2, \ldots$ which form an optimal sequence starting at k_1, i.e.

$$v_i \geq a_{ik_1} + v_{k_1} \qquad k_1 \in S_i$$

where S_i is the set of states that follows next in the chain passing through i. Since the "$=$" sign applies for $k_1 = j_1$ —see (2)—we have

$$(1) \qquad \qquad v_i = \underset{k \in S_i}{\text{Max}} (a_{ik} + v_k)$$

for all states which are not terminal ones. For terminal states we have

$$v_i = a_i$$

where a_i is the terminal payoff. This was the rule used in finding the value function for all nodes of the decision tree.

Equation (1)—the famous principle of optimality—will be discussed in detail below. It is the focal point of the DP approach.

Classification of decisions. For simplicity the number of states and the number of actions will be assumed finite at first. However, the horizon may be infinite. In that case the object of the decision process is considered to be the maximization of the sums of payoffs properly discounted. This is the subject of PART ONE and the foundation for everything else.

The decisions themselves and the sequence as well may be subject to risk and/or uncertainty. Risk, as usual, denotes the case of known probabilities. Uncertainty refers to the situation where a priori (subjective) probabilities are revised in the light of experience. Thus a Bayesian approach will be taken to uncertainty. We take it for granted that in all situations involving risk and uncertainty the relevant objective function is the expected value of utility [VON NEUMANN — MORGENSTERN]. These decision problems are usually cast in terms of Markov chains. They are the subject of PART TWO.

In certain problems of great practical interest the set of states and/or decision alternatives is continuous. Greater care is then needed in the application of the principle of optimality, but stronger properties of the value function may also be established (PART THREE).

Finally, the sequence of decisions need not be discrete but may take the form of a decision process in continuous time. In these problems DP has many points of contact with stochastic processes in continuous time, control theory, and the calculus of variations (PART FOUR).

A special class of decision sequences are games (not treated as such in this book). In multi-person games the person of the decision maker and hence the object of the decision will depend on the state of the game according to the rules of the game. If the strategies of all but one player may be considered given, then the game has been reduced for this given player to a sequence of decisions whose outcomes are known with certainty (case of pure strategies of other players and no chance moves) or risk, and where the object of decisions is maximization of an (expected) terminal payoff.

However, the restrictions imposed on decisions by information sets has made the application of the principle of optimality impractical, except for the simplest types of games (e.g. games with perfect information, see below page 14). But in game theory the sequential mode of analysis is not the prevailing one.

The recursive approach may be used also for games played in sequence rather than the moves of a single game [EVERETT].

General assumptions. Unless specified otherwise the decisions to be analysed will be considered to be sequential in time. The intervals between decisions are assumed to be of equal length. A constant discount factor ρ will be applied in each period, $0 < \rho < 1$. The discounting procedure needs no justification in economic analysis. It expresses both the terms of exchange of future against present goods and the chance of termination of the decision sequence. In addition—as BLACKWELL has emphasized [1965]—discounting proves of great advantage in the mathematical analysis of sequential problems since it defines a much smoother termination procedure than the fixed horizon—no discounting assumption.

§ 2. Geometric Interpretation

We begin with a detailed study on the simplest case of a Dynamic Program. Let there be a finite number of states. In each state there is a choice among a finite number of alternative actions. These actions will lead to new states, but in no case back will they lead to the state itself or to a state previously realized. The decision problem is terminated when any state in the terminal set is reached, which (in the view of these assumptions) must happen after a finite number of decisions.

We assume that payoffs are functions of transitions only i.e. of the state one is in and the state one goes to. Suppose now that there are several actions leading from one state to another state but having different payoffs. Then we can exclude as inefficient all actions with a smaller than the maximum payoff and retain only one with a maximal payoff. In this way a one-to-one correspondence is set up in each state between actions and the states they lead to. A geometric interpretation can now be given to the sequential decision problem. Let the states of the world—i. e., of a system—be viewed as vertices of a network. The possible actions, i. e., transitions designate edges of the network. The length of the edges are the payoffs—provided we admit, if necessary—negative measures of length (Fig. 1).

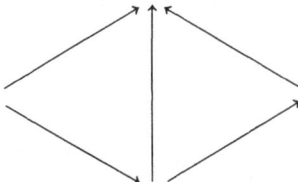

Fig. 1. Decision network

The geometric equivalent of a sequential decision problem is thus a directed graph whose nodes are the states and whose directed arcs are the decisions connecting the states. By assumption no closed cycles may occur, in particular at most one directed arc may occur between any pair of points. It is apparent that among others all decision trees are in this class of graphs. If payoffs (taken negative) are identified with arc length—admitting negative length—then the Dynamic Program may be interpreted as one of finding a shortest path in the directed network from an initial state i to a terminal state k, $k \in T$.

If in this network representation of the problem the lengths of all arcs turn out to be positive, then the prohibition on closed paths may be dropped from our assumptions since then a shortest path never involves closed cycles, so that states will not be repeated anyway. Arc lengths (possibly different for different directions) may now be defined for every pair of points and every direction, and the terminal set may be any point or set of points. We assume that the network is connected. (Otherwise the problem breaks up into disconnected i.e. mutually unrelated subproblems.)

The shortest path in a network. Fig. 2 shows the shortest path and the distances along these paths from various cities to the city of Bonn.

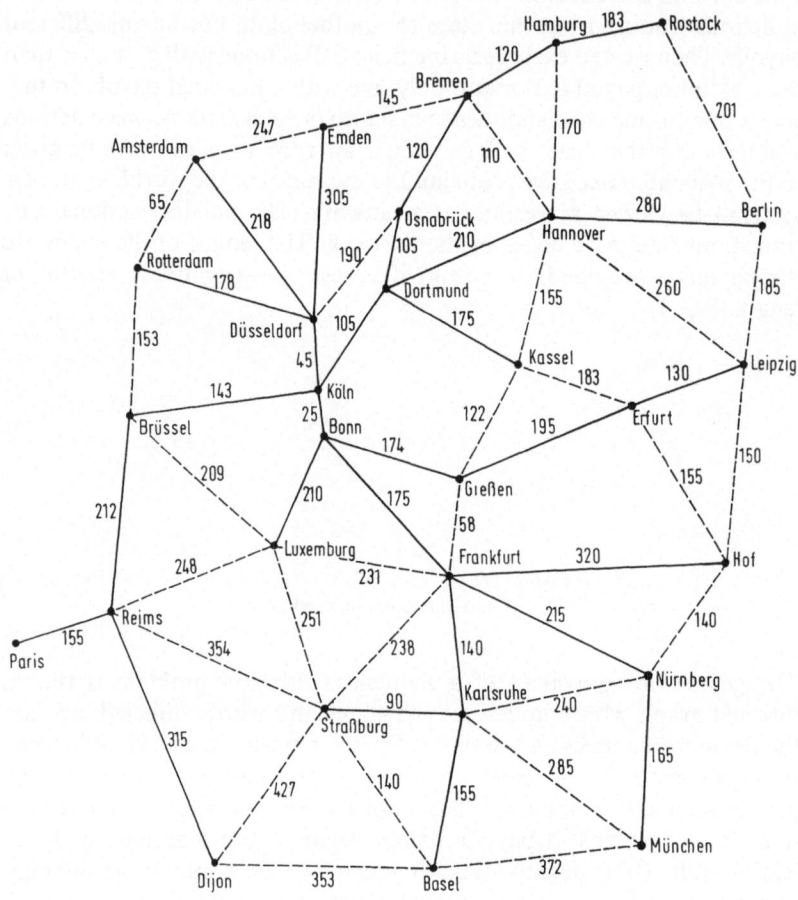

Fig. 2. Shortest routes to Bonn

The determination of the shortest path in a network is clearly sequential in nature since at each junction point in a network a new decision must be made as to which way to turn. (Road signs which are supposed to communicate the optimal decision to the traveller are often of little help.)

It is well known that the shortest path through a network may also be found by linear programming [DANTZIG (1963)]. We shall demonstrate below a resulting connection between the linear programming and dynamic programming approaches to the finite state decision problem.

The class of finite sequential problems without repetition includes many combinatorial problems, e.g. the sequencing of jobs on machines. A somewhat simper example—the allocation of indivisible resources (planes) is described below (§ 3).

Repetition of states. Let us now drop the restriction that no state may be repeated in a decision sequence and let instead a horizon n be fixed. Let payoffs a_{ij} be independent of the number of previous (of future realizations) of a transition. (Terminal payoffs are not excluded by this.)

Interpreting payoffs as lengths of arcs this decision problem has the following geometric representation. Find a path of maximal length from a given starting point traversing n edges without restriction as to the number of times any particular edge is traversed. Under our restrictions to a finite numbers of states, all problems with sufficiently large horizon must involve repetition of states.

Example: Optimal routing of a tramp ship.

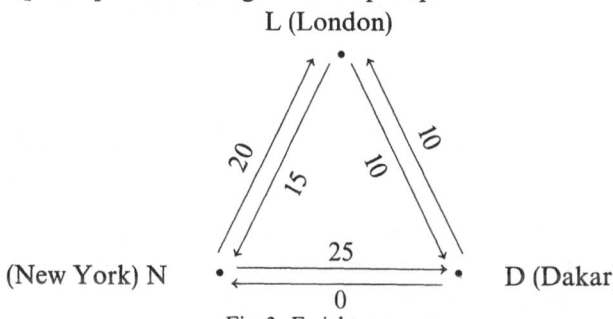

Fig. 3. Freight revenues

A freighter can make n trips between New York (N), London (L), and Dakar (D), say, before going into dry dock in London. The trips between any pair of these three ports are supposed to take the same time and to yield the freight revenues indicated in Fig. 3. Table 1 gives the optimal first destination for various starting points and various horizons n.

Table 1. *First destinations for various origins and horizons*

		horizon n					
		1	2	3	4	5	6
	D	L	N	L	L	L	L
location	L	—	N	N	N	N	N
	N	L	D	L	D/L	L	D/L

If $n > 2$ the optimum path contains the cycle N–L–N repeated as often as is feasible. Note however that initial (and terminal) decisions depend on the horizon n.

At this stage we may conjecture that all decision problems with repetition of states and long horizons have as an optimal solution a single optimal cycle repeated as often as is feasible.

§ 3. Principle of Optimality

A sequential decision problem may be decomposed into two parts

the first decision, and

the remaining ones.

Let i be the current state, and let $i_1, \ldots, i_t$ be the subsequent states traversed by an optimal decision sequence where i_t is a terminating state. The total payoff (here taken without discounting) v is clearly a function of the initial state i

$$v = v(i).$$

This value function is defined by the condition

$$(1) \qquad v(i) = \underset{i_1, i_2, \ldots, i_t}{\text{Max}} (a_{ii_1} + a_{i_1 i_2} + \cdots + a_{i_{t-1} i_t})$$

where the transitions $i_k i_{k+1}$ must be feasible. Feasibility may be described in terms of neighborhood sets S_i, i.e. the sets of points connected by an edge to i (see above page 5).

$$ij \quad \text{is feasible if} \quad j \in S_i.$$

The decision problem is feasible and a value function exists, provided a path exists from the initial state i to a terminating state i_t. Now,

$$v(i) = \underset{i_1 \in S_i}{\text{Max}} (a_{ii_1} + \underset{i_2, i_3, \ldots, i_t}{\text{Max}} (a_{i_1, i_2} + a_{i_2, i_3} + \cdots + a_{i_{t-1} i_t})).$$

By definition of the value function the second maximum term is $v(i_1)$, so that

$$(2) \qquad v(i) = \underset{i_1 \in S_i}{\text{Max}} [a_{ii_1} + v(i_1)].$$

Let the terminal payoff in a terminating state i_t be

$$(3) \qquad v(i_t) = a_{i_t}.$$

Starting with any terminating state i_t and working backwards by considering first states with a direct link to terminating states and then states linked to those, etc., the value function $v(i)$ may be found inductively for all states i of the given finite set.

The induction equation (2) is BELLMAN's Principle of Optimality [BELLMAN (1957)]. It introduces a value function or "state evaluation function" [RADNER (1963)] as a principal device for the recursive determination of solutions. The solution is a sequence of decisions which are to be taken in the various states. This solution may be expressed in terms of a decision rule

$$i_{k+1} = d(i_k) \quad \text{or, generally,}$$

$$j = d(i)$$

where i refers to the state in which the decision is made. The set of decision rules is denoted

$$\delta = \{d(i)\}.$$

Of course, value functions may be constructed for any given decision rule and such a construction is part of a method of solution known as policy iteration (see § 5). By recursion

(4) $$v(i,\delta) = a_{id(i)} + v(d(i),\delta)$$

where $v(i,\delta)$ denotes the value function under decision rule δ.

Example [BECKMANN/LADERMAN (1956)]: Plane Assignment.

As an illustration of the principle of optimality consider the problem of finding the best combination of two indivisible resources to meet a given demand.

Let the demand be a number of passengers and the resources be two types of planes

Plane	Capacity
DC 3	38
DC 6	58

Let the cost of operating a DC 6 on a given flight be 1.4 times that of running a DC 3. For any number n of passengers up to $n = 200$ it is desired to find the cheapest combination of planes that will carry them.

From 1 to 38 passengers are carried most cheaply by one DC 3; from 38 to 58 passengers by one DC 6.

To decide which is the cheapest cost of transporting 59 passengers we denote the minimum cost of transporting n passengers by $v(i)$ and have the recursive relation

$$v(i) = \text{Min} \left[v(i-58) + 1.4, v(i-38) + 1 \right] \quad \text{in particular}$$
$$v(59) = \text{Min} \left[v(1) + 1.4, v(21) + 1 \right]$$

where "Min" means the smaller of the two values in the brackets.

Since $v(1) = v(21) = 1$ one has $v(59) = 2$.

Now the minimum cost of carrying i passengers must be a non-decreasing function of i. (For one could always choose more space than necessary if this would decrease cost.) The minimum cost of carrying 59 passengers must therefore also be the minimum cost of carrying from 59 to 76 passengers (the full capacity load of two DC 3). It is clear therefore that one need consider only certain critical passenger numbers, namely those that would fill the various combinations of planes to capacity. In the case of two types of planes these are easily enumerated as in the following scheme

Table 1

DC 6 \ DC 3	0	1	2	3	4	5	6
0	0	38	76	114	152	190	228
1	58	96	134	172	210		
2	116	154	192	230			
3	174	212					
4	232						

Table 2

n	38	58	76	96	**114**	116	134	**152**	154	**172**	174
$c(i)$	1	1.4	2	2.4	3	2.8	3.4	4	3.8	4.4	4.2
$v(i)$	1	1.4	2	2.4	2.8	2.8	3.4	3.8	3.8	4.2	4.2

	190	192	**210**	212	**228**	**230**	232
$c(i)$	5	4.8	5.4	5.2	6	5.8	5.6
$v(i)$	4.8	4.8	5.2	5.2	5.6	5.6	5.6

In Table 2 we have listed the numbers of Table 1 in increasing order (row 1). The second row gives the cost that would arise if the combinations of DC 3 and DC 6 for which i is a full load were actually used.

In the third row the least cost is obtained by entering the smallest $c(i)$ for this or any larger i found in row 2. To summarize, all least cost combinations are also least cost combinations for full plane loads, but not vice versa, and Table 2 shows which full load combinations are inefficient (bold-face type).

Example: Shortest path.

The principle of optimality will now be applied to the shortest route in a network. For every pair of points i, j connected by a direct link (or road) ij let d_{ij} denote its length. (If there is no direct road from i to j we set d_{ij} infinite or consider it undefined.) S_i denotes again the set of points j to which a direct road from i exists. Consider the problem of shortest routes to destination $i=0$. The value function should now indicate the shortest distance from any given point to destination zero. In particular

$$D_0 = 0.$$

For any point j with a direct road to 0 one has

$$D_j = d_{j0}$$

(provided no shorter indirect path—to be discovered below—exists). For points with a direct connection to these points we have

$$(5) \qquad D_i = \underset{j \in S_i}{\text{Min}} \, (d_{ij} + D_j).$$

Notice that this is in fact the general principle for determining shortest paths and that it applies in this form regardless of the number of links that have yet to be traversed to the destination, and hence regardless of whether indirect paths turn out to be shorter than direct roads. It is also a check on any distance functions found in earlier rounds.

It may be instructive to consider the problem of shortest paths (in terms of travel time) when several modes m of travel are available on certain routes and certain transfer possibilities exist. Let

$$d_{ij}^m \qquad j, m \in S_i$$

indicate the travel time from i to j on mode m if such a combination exists at i $(j, m \in S_i)$.

Let $$t_i^{mn} \qquad m, n \in R_i$$

be the time involved in a transfer from mode m to mode n provided such a possibility exists at i. Let all trips be terminated at location zero in mode zero. Then minimal travel time through an optimal combination of modes and routes is dependent on both the point at which and on the mode on which the traveller is

$$D = D_i^m.$$

The principle of optimality assumes the form

(6)
$$D_0^m = t_i^{m0},$$

$$D_i^m = \underset{\substack{m,n \in R_i \\ n, j \in S_i}}{\text{Min}} \left[t_i^{mn} + d_{ij}^n + D_j^n \right].$$

Of course, through superposition of the several modal networks with their transfer points a single but larger network could have been constructed and the problem be reduced to the previous format. But the schema (6) is sometimes more transparent.

Games with perfect information. The principle of optimality may also be applied to max min decisions. Consider two-person-zero-sum-games with perfect information. In games of perfect information there exists a well defined state i of the game known to all participants. Consider two-person-zero-sum-games. Let $v(i)$ be the value of the game to player one, i. e., the expected payoff to player one if both follow optimal strategies *from now on*.

In the following we consider value functions only for those states i in which it is the first player's move. A move will be identified with the state of the game attained by that move. Let S_i be the set of states player one, can move to and let $j \in S_i$ be the move chosen. Let R_j denote the set of moves for the second player. His move will be one $\hat{k}$ for which

$$v(\hat{k}) = \underset{k \in R_j}{\text{Min}}\ v(k).$$

Player one will therefore choose j so as to obtain

$$\underset{j \in S_i}{\text{Max}}\ \underset{k \in R_j}{\text{Min}}\ v(k)$$

and this determines $v(i)$ recursively

(7)
$$v(i) = \underset{j \in S_i}{\text{Max}}\ \underset{k \in R_j}{\text{Min}}\ v(k) \qquad i \notin T$$

$$v(i) = a_i \qquad\qquad\qquad\qquad i \in T$$

where T is the set of terminal states for the game.

By means of this recursive principle it may be shown that in a game of perfect information no mixed strategies are required and that there exists, at each stage, a well determined value of the game. This formula also tells the best way of playing a game from any state on regardless of any mistakes that were made in the past moves, and it determines the allocation of the prizes when the game is terminated prior to reaching a terminal state. Of course, the actual calculation of the value function in real games such as chess and chequers is beyond our reach at this state of computer technology. However, certain end games in chess may be analysed in this manner [BELLMAN (1965)].

Repeated States. Consider now the case that all chains of decisions starting at i have equal length n. Let the possible transitions from i to j, the set S_i, and the payoffs a_{ij} all be independent of n. The value function depends now on both i and n. Let $v_n(i)$ be the value of state i with respect to decision chains of length n. The principle of optimality assumes the form

$$v_0(i) = a_i,$$

(8) $$v_n(i) = \underset{j \in S_i}{\mathrm{Max}} \left[a_{ij} + v_{n-1}(j) \right] \quad n = 1, 2, \ldots$$

Here a_i denotes the terminal payoff in state i.

Notice that while n is being increased the horizon is being extended through our moving backwards in time from the terminating decision to an earlier and earlier start of the decision process. This backward or recursive mode of induction is typical not only of DP but of much of mathematical problem solving [NEMHAUSER, p. 19—20], [POLYA (1957)].

The number of sequences of n decisions starting at i is finite even when repetition of states is permitted. Therefore the maximal sum of payoffs is well defined. The equation remains valid when state i occurs again in the decision chain. Through successive application of equation (8) one obtains then an implicit equation in the value function of the repeated state. Due to the finiteness of the decision process it is necessary to distinguish between the various times that a state is reached, e. g. by counting the remaining decisions n. For $v_n(i)$ depends on n, and in general so does the decision $j = d_n(i)$.

For the calculation of successive $v_n(i)$ by means of (8) it is of crucial importance that at every stage, in addition to the givens of the problem only $v_{n-1}(i)$ is required and need be stored in the memory and that the decision program is the same at every stage.

§ 4. Value Functions for Infinite Horizons: Value Iteration

Of great practical and theoretical interest is the case that while the number of states and of decisions remains finite the states may be traversed an infinite number of times. It is then imperative to discount future returns. We consider a discounted decision problem and the associated principle of optimality first in the finite case

(1) $$v_0(i) = a_i,$$

$$v_n(i) = \underset{j \in S_i}{\mathrm{Max}} \left[a_{ij} + \rho v_{n-1}(j) \right].$$

Now the question arises whether the decision problem remains meaningful when the horizon is extended to infinity, i. e., whether an

optimal solution exists and whether a value function may be associated
with it as before. Furthermore we must ask whether

$$\lim_{n \to \infty} v_n(i)$$

exists for all i, and whether it represents the value function of the problem
with infinite horizon. The following intuitive considerations may be
useful. Since the remaining decisions are not counted anymore the decision problem repeats itself at each passage through a state. From this it
follows that if an optimal decision exists it must have the following
structure: let k be the first state in a chain which is repeated. Then the
optimal chain will consist of an initial sequence from the initial state to
state k which is the first to be repeated and of infinitely many repetitions
of this cycle from k to k. Clearly there are only finitely many different
cycles from any state j to itself since no state may be repeated in a basic
cycle. Let

$$v^m(j)$$

be the maximum return for cycles of length m from j to j. Clearly a maximum of this function over the finite sets of m and j must exist. Moreover a
maximum yield per cycle and also a maximum yield $\bar{v}$ per decision must
exist defined by

$$\bar{v} = \frac{v^m(j)}{m}$$

as the sum of payoffs divided by the number of decisions. The solution
to the decision problem will consist in finding that cycle for which the
return $\bar{v}$ per decision is highest and in finding an initial transition to this
cycle.

A more rigorous analysis may now be given as follows. Note first
that in problems of fixed horizon (including an infinite horizon) a constant term may be added to each payoff without changing the solution.
Assume that this has been done so as to make all payoffs positive $a_{ij} > 0$.
Consider discounted sums of payoffs

$$a_{i_0 i_1} + \rho\, a_{i_1 i_2} + \cdots + \rho^{n-1} a_{i_{n-1} i_n}.$$

If the horizon is increased by one, a positive term $\rho^n a_{i_n i_{n+1}}$ is added.
Therefore

$$\sum_{k=0}^{n} \rho^k a_{i_k i_{k+1}} < \sum_{k=0}^{n+1} \rho^k a_{i_k i_{k+1}}$$

and this inequality is preserved after maximizing with respect to
$i_k, k = 1, \ldots, n$ on both sides and $k = n+1$ on the right hand side. Therefore

$$v_n(i) < v_{n+1}(i).$$

(In the presence of terminal payoffs this inequality is proved by backward induction.)

Let $\operatorname*{Max}_{ij} a_{ij} = a$. Then

$$v_n(i) \le a(1+\rho+\cdots+\rho^{n-1}) = a\frac{1-\rho^n}{1-\rho} < \frac{a}{1-\rho}.$$

For given i, $v_n(i)$ is therefore a bounded monotone sequence. Therefore its limit exists

(2) $$\lim_{n \to \infty} v_n(i) = v(i) \quad \text{(say)}.$$

We have shown that for infinite horizons the value function remains finite for every initial state i and approaches a limit $v(i)$. The decision problem for an infinite horizon is therefore well defined.

Does $v(i)$ satisfy the principle of optimality? Consider the principle of optimality (1) applied to finite horizons and restated to express maximization in terms of inequalities

(3) $$v_n(i) \ge a_{ij} + \rho v_{n-1}(j),$$
$$\text{``}=\text{'' for some } j \in S_i.$$

Since the limits of both sides exist as n goes to infinity they must satisfy the inequality and equation

$$v(i) \ge a_{ij} + \rho v(j),$$
$$\text{``}=\text{'' for some } j \in S_i$$

so that

(4) $$v(i) = \operatorname*{Max}_{j \in S_i} [a_{ij} + \rho v(j)].$$

Moreover the solution of (4) which we have just constructed is unique. To show this, let $v(i)$ and $u(i)$ be two solutions and let $\hat{j}$ be the maximizer of $v(i)$. Then

(5) $$v(i) = a_{ij} + \rho v(\hat{j})$$

and by definition of a maximum

$$u(i) = \operatorname*{Max}_{j} [a_{ij} + \rho u(j)]$$

(6) $$\le a_{ij} + \rho u(\hat{j}).$$

Subtracting (6) from (5)

$$v(i) - u(i) \ge \rho [v(\hat{j}) - u(\hat{j})].$$

Repeating the argument for $i=\hat{j}$ and substituting we obtain

$$v(i)-u(i)\geq\rho^{2}\left[v(\hat{\hat{j}})-u(\hat{\hat{j}})\right].$$

Continuing in this way and observing that u and v are bounded on the finite set $i=1,\ldots,m$ we see that the right hand side can be made arbitrarily small in absolute value. Therefore

$$v(i)-u(i)\geq 0.$$

Exchanging in the roles of u and v one shows similarly

$$u(i)-v(i)\geq 0 \quad \text{so that}$$

$$u(i)\equiv v(i) \qquad \text{QED}.$$

We have shown.

Theorem 1: *For every discounted infinite horizon decision problem with stationary finite alternatives there exists an associated value function satisfying the principle of optimality. It is the limit of value functions for finite horizon decision problems, and the unique solution of* (4).

The process of determining the value function as a limit of (1), i. e., through repeated application of the induction step (1), is known as value iteration. It represents one straightforward attack on the infinite horizon decision problem.

But does a value function of an infinite horizon problem always represent a solution? This is assured by

Theorem 2: *Every value function satisfying* (4) *determines an optimal decision rule which solves the infinite horizon decision problem.*

Proof: Suppose that $v(i)$ is a solution of (4). Let $\hat{j}$ be the maximizer of the bracket on the right hand side. Through successive substitution of (4) for $v(i)$, the right hand side of (4) becomes

$$v(i)=\underset{i_1}{\text{Max}}\left[a_{ii_1}+\rho\underset{i_2}{\text{Max}}\left[a_{i_1i_2}+\cdots+\rho\underset{i_n}{\text{Max}}\left[a_{i_{n-1}i_n}+\rho v(i_n)\right]\right]\right]$$

$$=\underset{i_1}{\text{Max}}\left[a_{ii_1}+\underset{i_2}{\text{Max}}\left[\rho a_{i_1i_2}+\cdots+\underset{i_n}{\text{Max}}\left[\rho^{n-1}a_{i_{n-1}i_n}+\rho^n v(i_n)\right]\right]\right].$$

Since $v(i)$ is bounded, the term $\rho^n v(i_n)$ can be made arbitrarily small by choosing n sufficiently large. Hence, $v(i)$ is seen to differ by arbitrarily little from the outcome of n maximizing actions. Since n is unlimited, $v(i)$ is shown to be the maximum discounted payoff sum of the decision problem with unlimited horizon.

§ 5. Policy Iteration

So far we have shown that a unique value function exists for the infinite horizon decision problem and that it generates an optimal decision rule. To show the complete correspondence of value functions and optimal decision rules it remains to go the reverse way: from optimal decision rules to the associated value function. This is the object of the present paragraph.

Theorem 1: *Every solution of (4.4) determines a decision rule which solves the infinite horizon decision problem.*

The proof is constructive and constitutes a practically useful algorithm, known as policy iteration.

Let an arbitrary decision rule

$$\delta_n = \{d_n(i)\}$$

be given. An associated value function v^n is then obtained by substituting this decision rule in the value determination equation (3.4)

(1) $$v^n(i) = a_{id(i)} + \rho v^n(d(i)).$$

This is a system of m linear equations in the m unkowns $v^n(i), i = 1, \ldots, m$. This equation system may be written

(2) $$v = a + \rho v Q$$

where $v = (v^n(1), v^n(2), \ldots, v^n(m))$, $a = (a_{1d(1)}, a_{2d(2)}, \ldots, a_{id(i)}, \ldots, a_{md(m)})$, and Q is a permutation matrix. The existence of a solution is seen by successive substitution.

(3) $$v = a + \rho Q a + \rho^2 Q^2 a + \cdots + \rho^n Q^n a + \ldots$$

and observing that the right hand side converges.

Now, if the decision rule δ_n is optimal then nothing more needs be proved, for (1) is then the principle of optimality. If the decision rule is not optimal then there exists an i_0 such that

(4) $$\underset{j \in S_{i_0}}{\text{Max}} [a_{i_0 j} + v(j, \delta)] > v(i_0, \delta).$$

Then define a new decision rule $\hat{\delta}$ by

$$\hat{d}(i) = \hat{j} \text{ where}$$

$$a_{ij} + v(\hat{j}, \delta) = \underset{j}{\text{Max}} [a_{ij} + v(j, \delta)].$$

We show that $v(i, \hat{\delta}) > v(i, \delta)$ for all i that are in the same recurrent chain (cycle) as i_0 under the new decision rule, and for i_0 itself if it is

2*

nonrecurrent under the new decision rule. In terms of the old value function we have under the new rule

$$v(i_0,\delta) < a_{i_0\,\hat{d}(i_0)} + \rho\, v(\hat{d}(i_0),\delta).$$

In terms of the new value function under the new rule

$$v(i_0,\hat{\delta}) = a_{i_0\,\hat{d}(i_0)} + \rho\, v(\hat{d}(i_0),\hat{\delta}).$$

Subtracting the second from the first

(5) $$v(i_0,\delta) - v(i,\hat{\delta}) < \rho\,[v(\hat{d}(i_0),\delta) - v(\hat{d}(i_0),\hat{\delta})].$$

The operation is now continued for $i=\hat{d}(i_0)$. The result corresponds to (3) with the $\leq$ sign replacing the $<$ sign. Upon multiplication with ρ on both sides we have

(6) $$\rho\,[v(\hat{d}(i_0),\delta) - v(\hat{d}(i_0),\hat{\delta})] \leq \rho^2\,[v(\hat{d}(\hat{d}(i_0)),\delta) - v(\hat{d}(\hat{d}(i_0)),\hat{\delta})].$$

If i_0 is recurrent under the new decision rule then by repeated application of (6) to $\hat{d}(\hat{d}(i_0))$, etc., we obtain a sequence of inequalities in which, finally at the kth step i_0 reappears on the right-hand side in

$$\rho^k\,[v(i_0,\delta) - v(i_0,\hat{\delta})].$$

Upon addition of all inequalities, terms for all states except i_0 of the recurrent chain cancel on both sides, thus yielding

$$v(i_0,\delta) - v(i_0,\hat{\delta}) < \rho^k\,[v(i_0,\delta) - v(i_0,\hat{\delta})]$$

or

$$(1-\rho^k)\,[v(i_0,\delta) - v(i_0,\hat{\delta})] < 0.$$

Since $\rho^k < 1$ we conclude that

$$v(i_0,\hat{\delta}) > v(i_0,\delta).$$

Applying (1) to the state preceding i_0 in the recurrent chain we obtain the same inequality for that state and, going backwards, for all states in the recurrent chain.

If i_0 is not recurrent repeated application of (6) as before shows that

$$v(i_0,\delta) \leq v(i_0,\hat{\delta})$$

for all successor states of i that are recurrent. Going backward to i_0 we finally come to (5) and obtain

$$v(i_0,\delta) < v(i_0,\hat{\delta}).$$

Thus, the algorithm raises the value function for at least one state while decreasing the value of no other state.

But since there are only finitely many different decision rules and hence finitely many linear equation systems for $v(i,\delta)$ and none may be repeated, the operation must terminate. This is possible only when (4) no longer holds, i. e., when for some N

$$v^N(i,\delta) \geqq a_{ij} + \rho v^N(j,\delta) \quad \text{for all } i.$$

By definition of $v^N(i)$ the "=" applies when $j = d^N(i)$ is substituted on the right hand side so that actually

$$(7) \qquad\qquad v^N(i) = \text{Max}_j [a_{ij} + \rho v^N(j)]$$

for all i. In this way it is shown that every optimal decision rule generates a value function satisfying the principle of optimality.

The method of policy iteration which constituted the constructive part of the proof will be discussed again in more detail for problems involving risk. The fundamental idea is to determine the value function associated with a given non-optimal decision rule. Such a value function can always be found by solving a system of linear equations. Although this value function is provisional it can serve nevertheless to indicate possible improvements in the decision rule. Thereupon, the value function is adjusted and the process is repeated. While each iterative step is fairly involved—requiring the solution of an equation system and a determination of maxima from finite sets—convergence is usually rapid. In any case only a finite number of repetitions is needed to determine the exact solution, while under value iteration the value function is only found approximately.

§ 6. Stability Properties

We have examined the infinite horizon problem as a limit of finite horizon problems. In practice the reverse is more common. The infinite horizon problem is a convenient vehicle for analysing the more complicated finite horizon problems of real life. Therefore the dependence of solutions on the horizon n is of interest. The same applies to the dependence on the discount factor. In this section we ask: How does the solution depend on the horizon and on the discount rate?

Theorem 1: *Let i be a state with a unique optimal initial decision j in the infinite horizon problem. For every fixed ρ, $0 < \rho < 1$, there exists an N such that the initial decision is identical for all $n > N$.*

Proof: The initial decisions in a problem with horizon $n+1$ is the maximizer of

$$a_{ij} + \rho v_n(j).$$

Define

$$\varepsilon = \min_{j \neq \hat{j}} [a_{ij} + \rho v(\hat{j}) - a_{ij} - \rho v(j)] > 0.$$

Since $v_n(i)$ converges for all i there exists an N such that

$$|v_n(i) - v(i)| < \frac{\varepsilon}{3} \quad \text{for all} \quad n > N.$$

Now

$$a_{ij} + \rho v_n(\hat{j}) - a_{ij} - \rho v_n(j) \geq a_{ij} + \rho v(\hat{j}) - \frac{\varepsilon}{3} - a_{ij} - \rho v(j) - \frac{\varepsilon}{3}$$

$$\geq \varepsilon - \frac{2}{3} \varepsilon = \frac{\varepsilon}{3} > 0.$$

Therefore for all $n > N$

(1) $$\max_j [a_{ij} + \rho v_n(j)] = a_{ij} + \rho v_n(\hat{j}).$$

Theorem 2: *Let n be fixed. Then the initial decision is the same in all problems with ρ from the interval $\rho_0 \leq \rho \leq 1$.*

Proof: For fixed finite n, $v_n(i)$ is a continuous function of ρ. Define

(2) $$\varepsilon = \min_{i, j \neq \hat{j}} [a_{ij} + \rho v_n(\hat{j}) - a_{ij} - \rho v_n(j)]$$

where $\hat{j}$ is the optimal decision for the decision problem with $\rho = 1$. Let the initial optimal decision be unique so that $\varepsilon > 0$. Since v_n is a continuous function of ρ, ρ_0 can be chosen so that $\varepsilon > 0$ in (2) for all $\rho_0 \leq \rho \leq 1$. This completes the proof.

When $\rho = 1$, Theorem 2 need not be true: It can happen that the first decision always depend on n. Example:

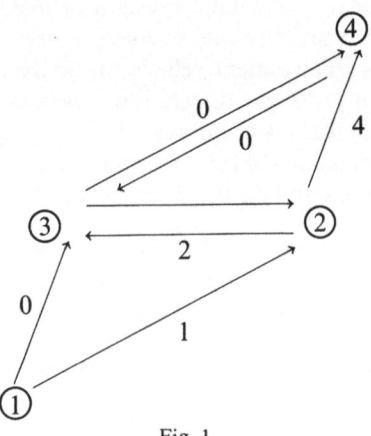

Fig. 1

One verifies that

$$d_1(1)=2$$
$$d_2(1)=2$$
$$d_3(1)=3$$
$$d_4(1)=2$$
$$d_5(1)=3 \text{ etc.}$$

$$d_n(1) = \begin{array}{ll} 3 & n \quad \text{odd}, \ n=3 \\ 2 & n \quad \text{even} \end{array}$$

However for large n the optimal cycle is independent of the horizon as we now show.

Nature of the optimal policy. If the horizon is unlimited, after a finite number of transitions some state must recur. The decision problem is then exactly the same as in the first occurrence of the state. Hence, this cycle must be repeated from then on forever. This must be the optimal policy also for sufficiently large n.

The value function for any state in the cycle is

$$v(i) = \sum_{r=1}^{m} a_{i_r i_{r+1}} \rho^{r-1} + \rho^m v(i)$$

where m is the length of the cycle or

$$v(i) = \frac{\displaystyle\sum_{r=1}^{m} a_{i_r i_{r+1}} \rho^{r-1}}{(1-\rho)\displaystyle\sum_{r=1}^{m} \rho^{r-1}}$$

since

$$\sum_{r=1}^{m} \rho^{r-1} = \frac{1-\rho^m}{1-\rho}.$$

Since $v(i)$ is maximal it follows that there exists no other closed cycle involving i which has a higher average return

$$\frac{\displaystyle\sum_r a_{i_r i_{r+1}} \rho^{r-1}}{\displaystyle\sum_r \rho^{r-1}} \quad \text{per transition.}$$

In the undiscounted problem for large n the optimal policy must involve the cycle highest average return

$$\text{Max} \frac{1}{m} \sum_{r=1}^{m} a_{i_r i_{r+1}}$$

except possibly at the beginning and at the end.

Now the highest average return over a cycle in finite network is a continuous function of ρ. It follows that for all ρ with $|1-\rho| < \varepsilon$ the optimal cycles of the discounted and of the undiscounted problem are identical. But the decision rules need not be the same since in the undiscounted case the initial decision may depend on the horizon as the example on page 22 shows. Note that the value function for the undiscounted finite horizon problem increases by an amount which is periodic with the length of the optimal cycle as period.

§ 7. Problems without Discount and with Infinite Horizon

How are the value functions of the discounted and of the undiscounted problem related? From the cyclic nature of the solution one has

$$(1) \qquad v(i,\rho) = c(i,\rho) + \frac{\rho^k}{1-\rho^m} \sum_{r=1}^{m} a_{i_r i_{r+1}} \rho^r$$

where $a_{i_r i_{r+1}}$ indicates the rth transition in the optimal cycle and $c(i,\rho)$ is a bounded term giving the value of the transition to the cycle, and k denotes the number of transitions before the cycle is reached. Now, as shown before, the optimal cycle is independent of ρ for all $\rho > \rho_0$. Consider

$$\lim_{\rho \to 1} (1-\rho) v(i,\rho) = \lim_{\rho \to 1} \frac{1-\rho}{1-\rho^m} \sum_{r=1}^{m} a_{i_r i_{r+1}} \rho^r = \frac{1}{m} \sum a_{i_r i_{r+1}} = \bar{v}$$

where $\bar{v}$ is the average payoff per decision. We have thus shown

Theorem: *For any sequential decision problem with finite alternatives in infinite horizon the following is true: For all states i*

$$(2) \qquad \lim_{\rho \to 1} (1-\rho) v(i,\rho) = \bar{v}$$

exists independent of i, and $\bar{v}$ is the maximum average payoff per decision in the undiscounted decision problem.

Consider now a new decision problem with payoff function

$$(3) \qquad \hat{a}_{ij} = a_{ij} - \bar{v}.$$

The new problem has a maximum average payoff of zero. Define a tentative value function $u(i)$ for the transformed problem as the solution of

$$u(i) = \underset{j \in S_i}{\text{Max}} \left[\hat{a}_{ij} + u(j) \right].$$

In terms of the original problem we have

$$u(i) = -\bar{v} + \underset{j \in S_i}{\text{Max}} \left[a_{ij} + u(j) \right] \quad \text{or}$$

(4) $$u(i) + \bar{v} = \underset{j \in S_i}{\text{Max}} \left[a_{ij} + u(j) \right].$$

The example of page 22 shows that even for the transformed problem a limit of value functions with finite horizon need not exist, since the value function may be periodic with respect to n. However, a solution $\bar{v}$, $u(j)$ of (4) always exists. To show this assume (without restriction) that all a_{ij} are positive. Let $j = d(i)$ be the optimal decision rule for the undiscounted average payoff. Apply the optimal decision rule to the states in the optimal cycle in succession

$$u(i) - u(d(i)) = a_{id(i)} - \bar{v}$$

$$u(d(i)) - u(d(d(i))) = a_{d(i)d(d(i))} - \bar{v} \qquad \text{etc.}$$

$$u(d^m(i)) - u(i) = a_{d^m(i)i} - \bar{v}.$$

Upon adding these equations the u terms cancel on both sides and we obtain

(5) $$\bar{v} = \frac{1}{m} \sum_{r=1}^{m} a_{i_r i_{r+1}}.$$

Choose an arbitrary $u(i)$, say $u(i) = 0$. From the first equation one obtains $u(d(i))$ and all other u's by solving the equations in succession. Repeating the argument of policy iteration it is shown that these $u(i)$ satisfy the principle of optimality and that the associated decision rule is optimal.

Theorem: *For an undiscounted decision problem with finitely many alternatives there exists an optimal average payoff $\bar{v}$ and a value function $u(i)$, satisfying (4), such that the optimal decision rule is the maximizer in (4).*

Note, that $u(i)$ is determined only up to an additive constant. Economically speaking, only the differences of these functions between states $u(i) - u(k)$ are well defined as indicating the advantage of beginning the decision in state i rather than state k — the total undiscounted value sums being infinite.

§ 8. Automobile Replacement

As an illustration consider the problem of an automobile owner who must decide every year whether to keep his old vehicle or to replace it by a different one. This decision will depend on the cost of a new automobile p_j, the trade-in value of his present car b_i, and on the expected operating cost c_i, for cars of various ages.

Table 1. *Purchase price, trade-in value and operating costs for automobiles of various ages*

Age i	Purchasing Price p_i	Trade-in Value b_i	Operating Costs c_i
0	2,200.—	1,800.—	220,—
1	1,500.—	1,250.—	260.—
2	1,100.—	910.—	305.—
3	800.—	680.—	355.—
4	640.—	530.—	410.—
5	560.—	465.—	470.—
6	500.—	400.—	540.—
7	450.—	360.—	630.—
8	410.—	335.—	750.—
9	370.—	210.—	910.—
10		100.—	

Table 1 presents data which are representative of United States automobiles. We assume that the age of operating automobiles is limited to nine years and also that there are no technical improvements so that the problem is stationary. Using the notation of Table 1 the principle of optimality may be stated as follows

$$
(1) \quad
\begin{aligned}
u_n(i) + \bar{v} &= \text{Min}\left[c_i + u_{n-1}(i+1),\ \text{Min}_j(p_j - b_i + c_j + u_{n-1}(j+1)) \right] \quad n > 0 \\
u_0(i) &= -b_i
\end{aligned}
$$

where the first Min operator refers to the two terms in the bracket.

Now, this problem is of a type where it makes sense to separate the costs of the decision a_{ij} (of reaching state j from i) from the cost of being in the next state a_j, say. In view of this, the principle of optimality can be reformulated in terms of the state attained by the decision j. This form is usually more convenient for policy iteration

(2) $$w_n(j) + \bar{w} = a_j + \operatorname*{Min}_i [a_{ji} + w_{n-1}(i)] \qquad n > 0$$

$$w_0(j) = a_j .$$

One advantage of this formulation is that the number of states after decisions is usually smaller than that of states in which decisions are made: The process is more controlled. In this particular example

(3) $$a_j = c_j \qquad\qquad\qquad\qquad\qquad \text{operating cost}$$

$$a_{ji} = \begin{cases} -b_{j+1} + p_i & \text{if} \quad i \neq j+1 \\ 0 & \text{if} \quad i = j+1 \end{cases} \qquad \text{trading costs} .$$

The principle of optimality has the form

(4) $$w_n(j) + \bar{w} = c_j + \operatorname*{Min}_i \left[w_{n-1}(j+1), \operatorname*{Min}_{i \neq j+1} (p_i - b_{j+1} + w_{n-1}(i)) \right] \qquad n > 0$$

$$w_0(j) = -b_j .$$

The principle of optimality states that one should not trade-in a vehicle unless the gain in value is equal to or more than the trading cost, and that the replacement should have that age for which purchasing price plus expected cost is minimal.

For the infinite horizon problem we merely drop the n subscripts on both sides of (4). Notice that the Min inside the bracket of (4) is then independent of i. If one trades it is always for a vehicle of the same – optimal – starting age i where i is the minimizer of $p_i + w(i)$. Equations (4) can be applied successively to determine w_n, and $\bar{w}$ is then the (uniquely determined) number for which this process converges. This is value iteration, but it is slow.

Instead let us apply policy iteration. Choose an initial arbitrary policy, say

> Trade any car six or more years old for a new car.
> Keep any car less than six years old.

(5) $$d_1(i) = \begin{cases} i & 0 \leq i < 6 \\ 0 & 6 \leq i \leq 9. \end{cases}$$

The associated average cost $\bar{w}^1$ and value function $w^1(i)$ are then determined as the solutions of the equation system (6)

(6) $$w^1(j) + \bar{w}^1 = c_j + w^1(j+1) \qquad\qquad j = 0, \ldots, 4$$

$$w^1(j) + \bar{w}^1 = c_j + p_0 - b_j + w^1(0) \qquad j = 5, \ldots, 9.$$

With the data of Table 1 we have in particular

$$w^1(0)+\bar{w}^1=220+w^1(1)$$
$$w^1(1)+\bar{w}^1=260+w^1(2)$$
$$w^1(2)+\bar{w}^1=305+w^1(3)$$
(7)
$$w^1(3)+\bar{w}^1=355+w^1(4)$$
$$w^1(4)+\bar{w}^1=410+w^1(5)$$
$$w^1(5)+\bar{w}^1=470+w^1(0)+2\,200-400.$$

Adding and noting that all $w^1(i)$ terms cancel we have

$$6\,\bar{w}^1=3\,820.—$$

$$\bar{w}^1=636.67\,.$$

Setting $w^1(0)=0$ we obtain by successive substitution in (7)

$$w^1(1)=416.67$$
$$w^1(2)=793.33$$
$$w^1(3)=1\,125.-$$
$$w^1(4)=1\,406.67$$
$$w^1(5)=1\,633.33\,.$$

Also

$$w^1(6)=1\,743.33$$
$$w^1(7)=1\,858.33$$
$$w^1(8)=2\,103.33$$
$$w^1(9)=2\,373.33\,.$$

To find an improved second decision rule we calculate the test variable

$$\underset{i}{\mathrm{Min}}\left[p_i+w^1(i)\right]$$

$$=\mathrm{Min}\,[2\,200,\ 1\,500+416.67,\ 1\,100+793.33,\ 800+1\,125,$$
$$640+1\,406.67,\ 560+1\,633.33,\ 500+1\,800,$$
$$450+1\,840,\ 410+1\,865,\ 370+1\,990]$$
$$=\mathrm{Min}\,[2\,200,\ 1\,916.7,\ 1\,893.3,\ 1\,925,\ 2\,046.7,\ 2\,193,$$
$$2\,300,\ 2\,290,\ 2\,275,\ 2\,360]$$
$$=1\,893.3=2+w^1(2)\,.$$

In deciding whether to trade, this test variable must be compared with

$$w^1(i) + b_i \qquad \text{for all } i$$

i	$w^1(i) + b_i$	$> 1\ 893.3$
0	1 800	no
1	1 666.67	no
2	1 703.33	no
3	180.5	no
4	1 936.67	yes
5	2 089.33	yes
6	2 143.33	yes
7	2 218.33	yes
8	2 438.33	yes
9	2 583.33	yes

The new decision rule is therefore

$$d_2(i) = \begin{cases} i & \text{for } i = 0, \dots, 3 \\ 2 & \text{for } i = 4, \dots, 9. \end{cases}$$

The equations determining $w^2(i)$ are

$$w^2(i) + \bar{w} = c_i + w^2(i+1) \qquad\qquad i = 0, \dots, 2$$
$$w^2(i) + \bar{w} = c_i + w^2(2) + p_2 - b_{i+1} \qquad i = 3, \dots, 9.$$

With particular values of Table 1

$$w^2(0) + \bar{w}^2 = 220 + w^2(1)$$
$$w^2(1) + \bar{w}^2 = 260 + w^2(2)$$
$$w^2(2) + \bar{w}^2 = 305 + w^2(3)$$
$$w^2(3) + \bar{w}^2 = 355 + 1\,100 - 530 + \bar{w}(2)$$
$$w^2(4) + \bar{w}^2 = 410 + 1\,100 - 465 + \bar{w}(2)$$
$$w^2(5) + \bar{w}^2 = 470 + 1\,100 - 400 + \bar{w}(2)$$
$$w^2(6) + \bar{w}^2 = 540 + 1\,100 - 360 + \bar{w}(2)$$
$$w^2(7) + \bar{w}^2 = 630 + 1\,100 - 335 + \bar{w}(2)$$
$$w^2(8) + \bar{w}^2 = 750 + 1\,100 - 210 + \bar{w}(2)$$
$$w^2(9) + \bar{w}^2 = 910 + 1\,100 - 100 + \bar{w}(2).$$

Adding the equations for $w^2(2)$ and $w^2(3)$ and cancelling terms we obtain

$$\bar{w}^2 = 615.$$

i	$w^2(i)$
0	—750
1	− 355
2	0
3	310
4	430
5	555
6	665
7	780
8	1 025
9	1 295

The test variable is

$$\operatorname*{Min}_{j}[p_j + w^2(j)] = 1070$$
$$= p_4 + w^2(4).$$

Comparing with $w^2(i) + b_i$.

i	$w^2(i) + b_i$	> 1070
0	1 050	no
1	895	no
2	910	no
3	990	no
4	960	no
5	1 020	no
6	1 065	no
7	1 140	yes
8	1 360	yes
9	1 505	yes

The third decision rule is therefore

$$d_3(i) = \begin{cases} i & i = 0, \dots, 6 \\ 4 & i = 7, 8, 9. \end{cases}$$

The equations determining $\bar{w}^3$ and $w^3(i)$ are

$$w^3(0)+\bar{w}^3=220+w^3(1)$$
$$w^3(1)+\bar{w}^3=260+w^3(2)$$
$$w^3(2)+\bar{w}^3=305+w^3(3)$$
$$w^3(3)+\bar{w}^3=355+w^3(4)$$
$$w^3(4)+\bar{w}^3=410+w^3(5)$$
$$w^3(5)+\bar{w}^3=470+w^3(6)$$
$$w^3(6)+\bar{w}^3=540+640-360+w^3(4)$$
$$w^3(7)+\bar{w}^3=630+640-335+w^3(4)$$
$$w^3(8)+\bar{w}^3=750+640-210+w^3(4)$$
$$w^3(9)+\bar{w}^3=910+640-100+w^3(4).$$

Adding the equations for $w^3(4)$, $w^3(5)$, $w^3(6)$ we obtain

$$\bar{w}^3=566 \ 2/3.$$

Setting $w^3(4)=0$ one obtains

	$p_i+w^3(i)$	$b_i+w^3(i)$	
$w^3(0)=-1\,126.67$	$1\,073.33$	673.33	$+$
$w^3(1)=-\ \ 780.—$	$720.—$	$470.—$	$-$
$w^3(2)=-\ \ 473.33$	626.67	436.67	$-$
$w^3(3)=-\ \ 211.67$	588.33	468.33	$-$
$w^3(4)=\quad\quad 0$	$640.—$	$530.—$	$-$
$w^3(5)=-\ \ 156.67$	716.67	621.67	$+$
$w^3(6)=-\ \ 253.33$		653.33	$+$
$w^3(7)=-\ \ 368.33$		728.33	$+$
$w^3(8)=-\ \ 613.33$		948.33	$+$
$w^3(9)=\quad 883.33$		$1\,093.33$	$+$

$$d^4(i)=\begin{cases} i & i=1,2,3,4 \\ 3 & i=0,5,6,7,8,9 \end{cases}$$

$$w^4(0)+\bar{w}^4 = -1\,800+800+355+w^4(4)$$
$$w^4(1)+\bar{w}^4 = \quad 260+w(2)$$
$$w^4(2)+\bar{w}^4 = \quad 305+w(3)$$
$$w^4(3)+\bar{w}^4 = \quad 355+w(4)$$
$$w^4(4)+\bar{w}^4 = \quad 410+800-465+w^4(3)$$
$$w^4(5)+\bar{w}^4 = \quad 470+800-400+w^4(3)$$
$$w^4(6)+\bar{w}^4 = \quad 540+800-360+w^4(3)$$
$$w^4(7)+\bar{w}^4 = \quad 630+800-335+w^4(3)$$
$$w^4(8)+\bar{w}^4 = \quad 750+800-210+w^4(3)$$
$$w^4(9)+\bar{w}^4 = \quad 910+800-100+w^4(3)$$
$$\bar{w}^4 = \quad 550.$$

Putting $w^4(3)=0$ one obtains

		$p_i+w^4(i)$	$b_i+w^3(i)$	
$w^4(0)= -980$			820	+
$w^4(1)= -520$			730	−
$w^4(2)= -245$		855	665	−
$w^4(3)= \quad 0$		800	680	−
$w^4(4)= \quad 195$		835	625	−
$w^4(5)= \quad 320$			785	−
$w^4(6)= \quad 430$			830	+
$w^4(7)= \quad 545$			905	+
$w^4(8)= \quad 790$			1\,125	+
$w^4(9)= 1\,060$			1\,270	+

$$d_5(i)=\begin{cases} i & i=1,2,3,4,5 \\ 3 & i=0,6,7,8,9. \end{cases}$$

The equations for the value functions are

$$w^5(0)+\bar{w}^5 = -1\,800+800+355+w^5(4)$$
$$w^5(1)+\bar{w}^5 = \quad 260+w^5(2)$$

$$w^5(2) + \bar{w}^5 = \quad 305 + w^5(3)$$

$$w^5(3) + \bar{w}^5 = \quad 355 + w^5(4)$$

$$w^5(4) + \bar{w}^5 = \quad 410 + w^5(5)$$

$$w^5(5) + \bar{w}^5 = \quad 470 + 800 - 400 + w^5(3)$$

$$w^5(6) + \bar{w}^5 = \quad 540 + 800 - 360 + w^5(3)$$

$$w^5(7) + \bar{w}^5 = \quad 630 + 800 - 335 + w^5(3)$$

$$w^5(8) + \bar{w}^5 = \quad 750 + 800 - 210 + w^5(3)$$

$$w^5(9) + \bar{w}^5 = \quad 910 + 800 - 100 + w^5(3)$$

$$\bar{w}^5 = \quad 545.$$

Setting $w^5(3) = 0$ one obtains the solution

$w^5(0) = -980$		820	+
$w^5(1) = -525$		725	−
$w^5(2) = -240$	860	660	−
$w^5(3) = 0$	800	680	−
$w^5(4) = 190$	830	620	−
$w^5(5) = 325$		790	−
$w^5(6) = 435$		835	+
$w^5(7) = 550$		910	+
$w^5(8) = 795$		1130	+
$w^5(9) = 1065$		1275	+

and so

$$d_6(i) = \begin{cases} i & i = 1,2,3,4,5 \\ 3 & i = 0,6,7,8,9. \end{cases}$$

Since $d_6(i) = d_5(i)$ policy iteration is ended and this is the optimal decision rule.

§ 9. Linear Programming and Dynamic Programming

The problem of determining a cycle which maximizes the average return per transition may also be cast as a linear program. Let x_{ij} denote the average number of transitions from i to j per period. (The reciprocal

$\dfrac{1}{x_{ij}}$ denotes the average time for the recurrence of this transition.) Total average payoff per period is then

(1) $$\sum_{ij} a_{ij} x_{ij} = \text{Max.}$$

The transitions are subject to the "flow constraint" that transitions to i and from i occur with equal frequency

(2) $$\sum_{j} x_{ij} = \sum_{k} x_{ki} .$$

Also there is a constraint on the total volume of flow: One transition per period

(3) $$\sum_{ij} x_{ij} = 1.$$

Moreover flows may not be negative

(4) $$x_{ij} \geq 0.$$

The linear program (1)–(4) has a feasible solution if the network structure allows any circular flow at all. Because of (3) there exists an upper bound A for the maximand

$$A = \underset{ij}{\text{Max}}\, a_{ij} .$$

Hence, there exists an optimal solution. With each constraint is associated a dual variable (or efficiency price), say, λ_i for (2) and μ for (3). The efficiency conditions are as follows: The Lagrange function

$$\sum_{ij} a_{ij} x_{ij} + \sum_{i} \lambda_i \sum_{j} (x_{ji} - x_{ij}) + \mu(1 - \sum_{ij} x_{ij})$$

has a maximum with respect to (x_{ij}) and a minimum with respect to (λ_i, μ) [BECKMANN (1959)]. Thus

(5) $$x_{ij} \begin{Bmatrix} \geq \\ = \end{Bmatrix} 0 \quad \text{according as} \quad a_{ij} - \lambda_i + \lambda_j - \mu \begin{Bmatrix} = \\ < \end{Bmatrix} 0.$$

It follows that

$$\lambda_i + \mu \geq a_{ij} + \lambda_j \quad \text{and that}$$

$$\text{``} = \text{''} \qquad \text{applies only when}$$

$$x_{ij} > 0.$$

Therefore

(6) $$\lambda_i + \mu = \underset{j \in S_i}{\mathrm{Max}}[a_{ij} + \lambda_j]$$

$$= a_{id(i)} + \lambda_{d(i)} \quad \text{(say)}.$$

In particular if for all i the Max of the right hand side of (6) is unique, then in view of (3)

$$x_{id(i)} = \frac{1}{k}$$

where k is the length of the cycle.

The positive flows x_{ij} of the linear programming solution thus indicate the optimal decision rules $j = d(i)$. The dual variables or efficiency prices represent the value function λ_i and average payoff μ respectively.

Linear Programming methods may therefore be used to solve the standard dynamic program with finite alternatives. In § 22 below the relationships between linear programs and dynamic programs will be further examined for some special problems.

Concluding remarks on discrete alternatives. A dynamic program can be formulated for a denumerably infinite number of states, but a solution need not exist. If it exists it need not involve any cycles. If the solution is cyclic, the problem may be reduced to one in terms of a finite number of states, without loss of efficiency, where the reduced set of states – which contains the cycle, of course, – will depend on the initial state.

Thus in the maintenance, knapsack, shortest path problems etc. arbitrarily large collections of ages, items, and nodes may be considered.

Even when all payoffs are bounded there need not exist a well defined solution even for a finite horizon (or for a single decision), and if one exists for every finite horizon it need not converge. These difficulties are to be kept in mind when trying to formulate dynamic programs with infinitely many alternatives, but they will hardly keep one from attacking in this way problems of meaningful empirical content.

References and Selected Reading to Part One

BECKMANN, M. J.: Lineare Planungsrechnung——Linear Programming. Ludwigshafen: Fachverlag für Wirtschaftstheorie und Ökonometrie 1959.
—, and J. LADERMAN: A Bound on the Use of Inefficient Indivisible Units. Naval Research Logistics Quarterly **3**, 245—252 (1956).
BELLMAN, R. E., and O. GROSS: Some Combinatorial Problems Arising in the Theory of Multistage Processes. SIAM **2**, 3, 175—183 (1954).
— Mathematical Aspects of Scheduling Theory. SIAM **4**, 3, 168—205 (1956).
— Notes on the Theory of Dynamic Programming IV——Maximization over Discrete Sets. Naval Research Logistics Quarterly **3**, 1+2, 67—70 (1956).

—, and S. E. DREYFUS: Applied Dynamic Programming. pp. 27—31. Princeton: Princeton University Press 1962.
— —Applied Dynamic Programming. pp. 229—231. Princeton: Princeton University Press 1962.
— On the Application of Dynamic Programming to the Determination of Optimal Play in Chess and Checkers. Proc. Nat. Acad. Sciences, USA. 53, 244—247 (1965).
— Dynamic Programming. Princeton: Princeton University Press 1957.
BLACKWELL, D.: Discrete Dynamic Programming. Ann. Math. Stat. 33, 2, 719—726 (1962).
— Discounted Dynamic Programming. Ann. Math. Stat. 36, 1, 226—235 (1965).
DANTZIG, G. B.: Linear Programming and Extensions. pp. 361—366. Princeton: Princeton University Press 1963.
EVERETT, H.: Recursive Games. In: Contributions to the Theory of Games III, eds.: M. Dresher, A. W. Tucker & P. Wolfe. pp. 47—78. Princeton: Princeton University Press 1957.
HELD, M., and R. M. KARP: A Dynamic Programming Approach to Sequencing Problems. SIAM 10, 196—210 (1962).
NEMHAUSER, G. L.: Introduction to Dynamic Programming. pp. 184—209. New York: Wiley 1966.
NEUMANN, J. VON, and O. MORGENSTERN: Theory of Games and Economic Behavior. 3 rd ed. Princeton: Princeton University Press 1953.
POLLACK, M., and W. WIEBENSON: Solutions of the Shortest-Route Problem. A Review. OR 8, 2, 224—230 (1960).
POLYA, G.: How to Solve it. New York: Doubleday & Company, Inc. 1957.
RADNER, R.: Notes on the Theory of Economic Planning. Athens: Center of Economic Research 1963.
SADOWSKI, W.: A Few Remarks on the Assortment Problem. MS 6, 1, 13—24 (1959).
SAVAGE, R.: Cycling. NRLQ 3, 3, 163—175 (1956).
WALD, A.: Statistical Decision Functions. New York: Wiley 1950.

PART TWO

Risk

§ 10. Basic Concepts

Classification. Uncertainty means that the decision maker does not know

> the state of the system, or
>
> the payoff of an action, or
>
> the new state generated by an action.

Risk means that he knows their probabilities.

The case of greatest practical importance is that where the decision maker knows the state of the system with certainty and knows the probabilities of payoff and transitions to new states for every actions.

Markov chains. When a system is described in terms of transition probabilities from states to states one speaks of a stochastic process. The process is Markovian when the transition probabilities depend only on the current state and not on past history. This assumption is invalidated for instance when the transition probabilities are still being estimated on the basis of past experiences, i. e., in the case of uncertainty (§ 18, below). Let there be a finite number of states $i = 1, \ldots, m$, and let transitions occur at intervals of fixed length and be Markovian. Then we have a Markov chain defined by the transition probabilities p_{ij} for transitions from i to j.

A Markov chain may be described graphically by a network whose nodes are the states and whose directed arcs are the possible transitions.

Example: Let a machine be in one of two states

> 1 operating or
>
> 0 failed.

When operating the machine will fail in the given period with probability p_{10}; when failed a successful repair will be made in the same period with probability p_{01}. Fig. 1 shows the graph of this process

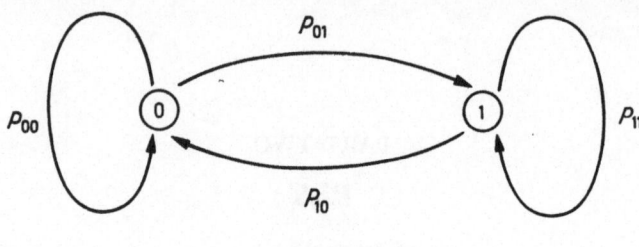

Fig. 1

The basic information about a Markov chain is the matrix of transition probabilities

$$P = ((p_{ij})).$$

The p_{ij} are subject to the following restrictions

(1) $p_{ij} \geq 0$

and

(2) $\sum_j p_{ij} = 1$ for all i

since the probabilities of possible transitions from any given state must add up to one. Because of restrictions (1) and (2) P is a stochastic matrix.

Higher order transition probabilities are obtained as follows. Let $p_{ij}^{(2)}$ be the probability of a transition from i to j in two steps

$$p_{ij}^{(2)} = \sum_r p_{ir} p_{rj} . \quad \text{Thus}$$

$$P^{(2)} = ((p_{ij}^{(2)})) = P^2 \quad \text{and generally}$$

(3) $P^{(n)} = ((p_{ij}^{(n)})) = P^n .$

For instance for the stochastic matrix

$$P = \begin{pmatrix} \frac{1}{2} & \frac{1}{2} \\ \frac{1}{10} & \frac{9}{10} \end{pmatrix} = \begin{pmatrix} 0.5 & 0.5 \\ 0.1 & 0.9 \end{pmatrix}$$

we have

$$P^2 = \begin{pmatrix} 0.30 & 0.70 \\ 0.14 & 0.86 \end{pmatrix}$$

$$P^3 = \begin{pmatrix} 0.220 & 0.780 \\ 0.156 & 0.844 \end{pmatrix}$$

$$P^4 = \begin{pmatrix} 0.1880 & 0.8120 \\ 0.1624 & 0.8376 \end{pmatrix}$$

$$P^5 = \begin{pmatrix} 0.17520 & 0.82480 \\ 0.16496 & 0.83504 \end{pmatrix}$$

$$P^6 = \begin{pmatrix} 0.170080 & 0.829920 \\ 0.165984 & 0.834016 \end{pmatrix}$$

$$P^7 = \begin{pmatrix} 0.1680320 & 0.8319680 \\ 0.1663936 & 0.8336064 \end{pmatrix}$$

$$P^8 = \begin{pmatrix} 0.16721280 & 0.83278720 \\ 0.16655744 & 0.83344256 \end{pmatrix}$$

$$\lim_{n \to \infty} P^n = \begin{pmatrix} \frac{1}{6} & \frac{5}{6} \\ \frac{1}{6} & \frac{5}{6} \end{pmatrix}.$$

Consider now the concept of state probabilities. Let

$$\pi_i^{(0)} \quad i = 1, \ldots, m$$

denote the probability that the system is initially in state i. In most applications to DP

$$\pi_i^{(0)} = \begin{cases} 1 & \text{for some } i \\ 0 & \text{for all others} \end{cases}$$

since the current state is assumed known with certainty. Let $\pi_i^{(n)}$ denote the probability that the system will be in state i n periods from now (given whatever we know about its present state). Clearly

$$\pi_j^{(1)} = \sum_i \pi_i^{(0)} p_{ij}$$

and generally

(4) $$\pi_j^{(n+1)} = \sum_i \pi_i^{(n)} p_{ij}.$$

The vector of state probabilities $\pi_i^{(n)}$ is written $\pi^{(n)}$. Equation (4) is a system of difference equations for $\pi^{(n)}$

$$\pi^{(n+1)} = \pi^{(n)} P.$$

The solutions of such difference equations have the form

$$\pi^{(n)} = q \ \lambda^n$$

where λ is a characteristic root of the matrix P and q the associated eigenvector

(5) $$\det(P-\lambda I)=0,$$

$$q\lambda=qP.$$

From restriction (2) it follows that

$$\lambda=1 \text{ is always a characteristic root.}$$

The general solution is a linear combination

(6) $$\pi^{(n)}=\sum_{r=1}^{m} q^{(r)} \lambda_r^n.$$

If $\lambda_1=1$ and all other roots λ_r satisfy the condition

$$|\lambda_r|<1 \quad r=2,\ldots,m$$

then

$$\lim_{n\to\infty} \pi^{(n)}=q^{(1)} \quad \text{exists}$$

$$=\pi \quad \text{say.}$$

This means that the state probabilities approach limiting values which we may call stationary state probabilities. Since the eigenvector $q^{(1)}$ of $\lambda=1$ is unique the stationary state probabilities are seen to be independent of the initial conditions. From (4) one has for the stationary state probabilities π

(7) $$\pi_j=\sum_i \pi_i p_{ij} \quad \text{or in matrix notation}$$

$$\pi=\pi P$$

which shows that $\pi=q^{(1)}$ is the row eigenvector of the eigenvalue $\lambda=1$.

Markov chains with this property are called ergodic. A test for ergodicity is that for some n

$$p_{ij}^{(n)}>0 \quad \text{for all } i \text{ and } j$$

i.e. that some P^n is a positive matrix, $P^n>0$, (so that in fact all $P^r>0$ $r\geqq n$).

Unless said otherwise we restrict ourselves to ergodic Markov chains. We note for later use that all components of π are positive

$$\pi_i>0 \quad i=1,\ldots,m.$$

This follows from the observation that

$$\pi = \pi P \quad \text{implies}$$

$$\pi = \pi P^n$$

and that $P^n > 0$ for some n.

Returns. Let now returns b_i, b_{ij} (measured in utility) be associated with states i and transitions ij. If the system is in state i there is an immediate return in this period whose expected value is

$$b_i + \sum_j p_{ij} b_{ij} = a_i \quad \text{say.}$$

The expected return n periods from now is

$$\sum_i \pi_i^{(n)} \left[b_i + \sum_j p_{ij} b_{ij} \right] = \sum_i \pi_i^n a_i$$

$$= \sum_i a_i \left[\sum_i q_i^{(r)} \lambda_r^n \right] \quad \text{using (6).}$$

As n increases this approaches

(8) $$\sum_i a_i q_i^{(1)} = \sum_i a_i \pi_i = a \quad \text{(say).}$$

In the ergodic case the expected return per period approaches a constant limiting value a which is independent of initial conditions.

§ 11. The Value Function

Consider next the expected discounted earnings over n periods given that the system is now in state i. This may be called a value function $v_n(i)$. Note however, that this value function does not depend on any deliberate action but reflects only the stochastic behavior of the system. Now

(1) $$v_0(i) = 0$$

(2) $$v_n(i) = a_i + \rho \sum_j p_{ij} a_j + \rho^2 \sum_j p_{ij}^{(2)} a_j + \cdots + \rho^n \sum_j p_{ij}^{(n)} a_j \quad n > 0$$

(3) $$= a_i + \rho \sum_j p_{ij} v_{n-1}(j)$$

in view of the definition of $v_n(i)$.

The recursive definition (1), (3) will play an important part later on.
Let A be an upper bound of the expected one period return a_i

$$a_i \leqq A.$$

From (2) an upper bound of the expected n period returns is obtained

$$v_n(i) \leq A + \rho \sum_j p_{ij} A + \rho^2 \sum_j p_{ij}^{(2)} A + \cdots + \rho^n \sum_j p_{ij}^{(n)} A$$

$$\leq A + \rho A + \rho^2 A + \cdots + \rho^n A < \frac{A}{1-\rho} \qquad 0 \leq \rho < 1.$$

Next suppose that all expected one period payoffs are positive $a_i > 0$. If they are not, a positive constant may be added to all returns without changing the nature of the problem (provided n is fixed). Consider

$$v_{n+1}(i) - v_n(i) = \rho^n \sum_j p_{ij}^{(n)} a_j > 0.$$

The sequence $v_n(i)$ is therefore monotone increasing and bounded. Hence its limit exists

$$\lim_{n \to \infty} v_n(i) = v(i) \qquad \text{say.}$$

For every state i there exists thus a well defined present value of the unlimited stream of future payoffs. Going to the limit in (3) one obtains

(4) $$v(i) = a_i + \rho \sum_j p_{ij} v(j).$$

Consider now the case of undiscounted future returns. From (2)

(5) $$v_n(i) = a_i + \sum_j p_{ij} a_j + \cdots + \sum_j p_{ij}^{(n)} a_j.$$

Now

$$p_{ij}^{(n)} = \sum_k \pi_k^{(n-1)} p_{kj}.$$

Substituting from (10.6)

$$= \sum_{k=1}^{m} \sum_{r=1}^{m} q_k^{(r)} \lambda_r^{n-1} p_{kj}.$$

Substituting in (5)

$$v_n(i) = a_i + \sum_j \sum_{t=1}^{n} \sum_{k=1}^{m} \sum_{r=1}^{m} q_k^{(r)} \lambda_r^{t-1} p_{kj} a_j.$$

Since

$$\lambda_1 = 1 \qquad |\lambda_r| < 1 \qquad r \neq 1$$

$$v_n(i) = a_i + n \sum_j \sum_{k=1}^{m} q_k^{(1)} p_{kj} a_j +$$

$$+ \sum_{j} \sum_{t=1}^{n} \sum_{k=1}^{m} \sum_{r=2}^{m} q_k^{(r)} \lambda_r^{t-1} p_{kj} a_j$$

(6)
$$= a_i + (n-1) \sum_{j} \sum_{k=1}^{m} \pi_k p_{kj} a_j + f_{i,n}(\lambda_2, \lambda_3, \ldots, \lambda_m)$$

where $f_{i,n}(\lambda_2, \ldots, \lambda_m)$ denotes the remaining terms. Observe that

(7)
$$\sum_{k=1}^{m} \pi_k p_{kj} = \pi_j$$

and that

(8)
$$\sum \pi_j a_j = a.$$

Substituting

(9)
$$v_n(i) = na + u_n(i)$$

where $u_n(i)$ consists of constant terms and of terms involving λ_r^t, $r \neq 1$, which approach a finite limit as $n \to \infty$

$$\lim_{n \to \infty} u_n(i) = u(i).$$

Asymptotically therefore

$$v_n(i) \cong na + u_i.$$

The undiscounted period return is asymptotically equal to a constant term u_i plus n times a stationary average return a per period.

Substitute (9) in the recursive equation for the value function (3) (noting that $\rho = 1$)

$$na + u_n(i) = a_i + \sum_{j} p_{ij}[(n-1)a + u_{n-1}(j)]$$

or

(10)
$$u_n(i) + a = a_i + \sum_{j} p_{ij} u_{n-1}(j)$$

with

$$u_0(i) = 0$$

since

$$v_0(i) = 0 \qquad \text{by (1)}.$$

The value function of the undiscounted problem may be determined recursively using (9) and (10) provided we know the stationary average return a per period.

Going to the limit in (10) we obtain

(11)
$$u(i) + a = a_i + \sum_{j} p_{ij} u(j).$$

From (11) the expression (10.8) for a may be obtained again. Multiplying (11) by π_i and summing

$$\sum_i \pi_i u_i + a \sum_i \pi_i = \sum_i \pi_i a_i + \sum_{ij} \pi_i p_{ij} u_j.$$

Noting

$$\sum_i \pi_i = 1 \quad \text{and} \quad \sum_i \pi_i p_{ij} = \pi_j$$

we have

$$a = \sum_i \pi_i a_i.$$

So far the value functions of the Markov chain with unlimited horizon have been shown to satisfy certain equations: (4) for the discounted and (11) for the undiscounted problem. The following question may now be asked:

Is every solution of (4) the value function of the Markov chain?

Our construction shows that solutions always exist. In the discounted case (4) the solution is unique. Suppose that $v(i)$ and $w(i)$ are two solutions. Substituting in (4) and subtracting we have

(12) $$v(i) - w(i) = \rho \sum_j p_{ij} [v(j) - w(j)].$$

Write $v(i) - w(i) = z(i)$.

Since $z(i)$ is defined on a finite set $i = 1, \ldots, m$, it is bounded

$$|z(i)| \leq c.$$

Applying (12) successively we have

$$z(i) = \rho^n \sum_j p_{ij}^{(n)} z(j)$$

$$z(i) \leq \rho^n c < \varepsilon \quad \text{for } n \text{ sufficiently large.}$$

In the same way one shows $z(i) \geq 0$ so that $z(i) = 0$. Therefore the solution of (4) is unique.

Consider now two solutions

$$u(i), a \quad \text{and} \quad w(i), b \quad i = 1, \ldots, m \quad \text{of (11).}$$

Substituting in (11) and subtracting we have

$$u(i) - w(i) + a - b = \sum_j p_{ij} [u(i) - w(i)]$$

or

$$z_i = \sum_j p_{ij} z_j + c \quad \text{where} \quad c = b - a$$

$$z_i = u_i - w_i.$$

Substituting successively

$$z_i = nc + \sum_j p_{ij}^{(n)} z_j$$

(13)
$$= nc + \sum_j \pi_j z_j + \varepsilon \qquad \text{for large } n$$

where ε may be made arbitrarily small,

$$= nc + k + \varepsilon \qquad k \text{ a constant.}$$

Since z_i is bounded this is a contradiction unless $c=0$. Therefore the stationary average return a is uniquely determined by (11).

Going to the limit in (13) we have

(14)
$$z_i = \sum_j \pi_j z_j \qquad \sum_j \pi_j = 1 \qquad \pi_j > 0$$

which has as its only solution

$$z_i = \text{constant.}$$

Therefore, the u_i are determined only up to an additive constant and may differ from the true value function

$$\lim_{n \to \infty} u_n(i)$$

by a constant.

§ 12. The Principle of Optimality

So far the stochastic process was allowed to run its course without interference. Let a decision maker be introduced who observes the state of the system and chooses an action. A decision plan in which the choice of an action depends on the state observed is called a strategy. It is fundamentally different from a decision plan for all periods in which actions are fixed in advance and are not conditional on future states.

In general, action k will influence both events—the transition probabilities and the payoffs. We must modify

$$p_{ij} \to p_{ij}^k$$
$$a_i \to a_i^k .$$

As soon as k is specified as a function of i by a decision rule

(1)
$$k = d(i)$$

we have once more a pure Markov chain. Assume it to be ergodic under all possible decision rules. Then our previous results apply. In particular with each decision rule $\delta = \{d(i)\}$ is associated a value function $v_n(i,\delta)$ in the discounted case and $u_n(i,\delta)$ plus a stationary average payoff $a(\delta)$ in the undiscounted case.

Optimal decisions. A sequence of optimal decisions is defined as one which maximizes the expected return over n periods given the present state of the system. In other words it maximizes the value function. Let $v_n(i)$ – without δ variable – denote the value function associated with a sequence of optimal decisions. By definition

$$(2) \qquad v_n(i) = \operatorname*{Max}_{k_1}\left[a_i^{k_1} + \rho \sum_j p_{ij}^{k_1} \operatorname*{Max}_{k_2}\left[a_j^{k_2} + \right.\right.$$

$$\left.\left. + \rho \sum_h p_{jh}^{k_2} \operatorname*{Max}_{k_3}\left[a_h^{k_3} + \cdots + \operatorname*{Max}_{k_n}\left[a_y^{k_n} + \rho \sum_z p_{yz}^{k_n} v_0(z)\right]\right]\right]\right]$$

or

$$(3) \qquad v_n(i) = \operatorname*{Max}_k\left[a_i^k + \rho \sum_j p_{ij}^k v_{n-1}(j)\right]$$

where the definition of $v_n(j)$ has been used to separate the decision chain into the first decision and the remaining ones. With

$$(4) \qquad\qquad v_0(i) = a_i \qquad a_i \text{ a terminal payoff}$$

or more simply

$$v_0(i) = 0.$$

(3) determines the value function inductively for decision chains of all finite lengths n. This inductive (or recursive) calculation of the value function is known as *value iteration*. Through value iteration the value function is uniquely determined for all i and n. The optimal decision rule is then obtained as the maximizer of the right hand side in (3). From the construction (2) it is clear that every decision rule obtained as the maximizer in (3) is optimal.

Since under our usual assumptions the value function is monotone and bounded its limit exists

$$\lim_{n \to \infty} v_n(i) = v(i)$$

and satisfies the principle of optimality

$$(5) \qquad\qquad v(i) = \operatorname*{Max}_k\left[a_i^k + \rho \sum_j p_{ij}^k v(j)\right].$$

Is every solution of (5) the value function of an optimal decision chain with infinite horizon? To prove that it is we show uniqueness of the solution of (5). Any solution must therefore agree with the one constructed above, viz. the value function for limits of optimal decision chains.

Let $v(i)$ and $w(i)$ be two solutions and let $\hat{k}$ be the maximizer in (5) using v. Then

$$v(i)=a_i^{\hat{k}}+\rho\sum_j p_{ij}^{\hat{k}}v(j)$$

$$=a_i^{\hat{k}}+\rho\sum_j \hat{p}_{ij}v(j), \quad \text{say.}$$

$$w(i)\geq a_i^{\hat{k}}+\rho\sum_j \hat{p}_{ij}w(j)$$

by definition of the maximum.

Subtracting and writing $v(i)-w(i)=z(i)$

$$z(i)\leq\rho\sum_j \hat{p}_{ij}z_j .$$

Through successive substitution

$$z(i)\leq\rho^n\sum_j \hat{p}_{ij}^{(n)}z_j .$$

Observe that $z(i)$ (being defined on a finite set) is bounded and letting $n\to\infty$ we conclude

$$z(i)\leq 0.$$

In the same way by considering the decision rule which is optimal for w we obtain

$$z(i)\geq 0 \quad \text{all } i$$

and so $v(i)\equiv w(i)$ QED.

Consider now the undiscounted decision problem. Equation (3) applies also when $\rho=1$. Writing the value function

$$v_n(i)=na+u_n(i) \quad \text{(cf. 11.9)}$$

one has the equivalent formulation

(6) $$u_n(i)+a=\underset{k}{\text{Max}}\left[a_i^k+\sum_j p_{ij}^k u_{n-1}(j)\right]$$

$$u_0(i)=0$$

which contains an arbitrary constant a. In § 11 it was shown that for any given fixed decision rule δ there exists a unique constant a such that $u_n(i)$ remains bounded as $n\to\infty$. If it can be shown that the decisions in the chain (2)

$$k_n=d(i,1), \quad k_{n-1}=d(i,2), \quad k_1=d(i,n)$$

converges as n increases

$$d(i,n)\to d(i)$$

the existence of a value function $u(i)$ and of an average return a for the undiscounted decision problem with infinite horizon would follow. It is more convenient to give a direct construction of the optimal decision rule for the infinite horizon problem and to determine the value function *pari passu*. This construction is called policy iteration and constitutes a powerful method of solving ergodic infinite horizon decision problems [HOWARD ch. 4].

§ 13. Policy Iteration

Consider a first decision rule

$$(1) \qquad\qquad k=d(i).$$

We assume that actions should depend only on the state of the system and not (say) on the number of previous decisions taken or on the initial state. It will be proved later that in the infinite horizon problem the optimal policy is always stationary, i. e., depends only on the current state. With every (stationary) decision rule $\delta = \{d(i)\}$ there is associated a value function, as shown in § 11, for

$$p_{ij}^{d(i)} = p_{ij}^*, \quad \text{(say)}$$

defines an ordinary Markov chain.

   This value function is obtained as a solution (unique in a, unique up to an additive constant for u_i) of the linear equation system

$$u(i)+a=a_i^{d(i)}+\sum_j p_{ij}^{d(i)} u(j)$$

$$(2) \qquad\qquad = a_i^* + \sum_j p_{ij}^* u(j), \quad \text{say.}$$

In terms of this function $u(i)$ determine the maximizer $k=\hat{k}$ of

$$\operatorname*{Max}_k \left[a_i^k + \sum_j p_{ij}^k u(j) \right]$$

$$(3) \qquad\qquad = a_i^{\hat{k}} + \sum_j p_{ij}^{\hat{k}} u(j)$$

$$= \hat{a}_i + \sum_j \hat{p}_{ij} u(j), \quad \text{say.}$$

It is clear that the maximizer does not depend on the arbitrary additive constant in $u(i)$. Associated with the decision rule

$$(4) \qquad\qquad d(i)=\hat{k}(i)$$

is a new value fuction $\hat{u}(i)$, say, defined as the solution of

$$(5) \qquad\qquad \hat{u}(i)+\hat{a}=\hat{a}_i+\sum_j \hat{p}_{ij}\hat{u}(j)$$

and also a new stationary average return $\hat{a}$.

From (2), (3) and the definition of a maximum it follows that

$$u(i)+a \leqq a_i^{\hat{k}}+\sum_j p_{ij}^{\hat{k}} u(j)$$

and $<$ for at least one i. This may be written

(6) $u(i)+a \leqq \hat{a}_i+\sum_j \hat{p}_{ij} u(j)$ and for at least some i.

Subtracting (6) from (5)

(7) $\hat{u}(i)-u(i)+\hat{a}-a \geqq \sum_j \hat{p}_{ij}\left[\hat{u}(j)-u(j)\right]$ for at least some i.

Let $\hat{\pi}_i \ (i=1,...,m)$ be the stationary state probabilities under the new decision rule $\hat{k}$. Multiplying (7) by $\hat{\pi}_i$ and summing

(8) $\sum_i \hat{\pi}_i(\hat{u}(i)-u(i))+\hat{a}-a > \sum_{i,j} \hat{\pi}_i \hat{p}_{ij}[\hat{u}(j)-u(j)]$

since all $\hat{\pi}_i$ are positive.

In view of

$$\sum_i \hat{\pi}_i \hat{p}_{ij}=\hat{\pi}_j$$

the first and last terms cancel so that

(9) $\hat{a}-a>0.$

Replacing a decision rule by one that maximizes returns in terms of the provisional value function $u(i)$ increases, therefore, the average return a. If further improvements are possible the process is repeated. Since no decision rule can occur twice the finite set of possible decision rules is eventually exhausted. Now the impossibility of further improvements means that under the last decision rule $\tilde{k}$

$$\tilde{u}(i)+\tilde{a}=a_i^{\tilde{k}}+\sum_j p_{ij}^{\tilde{k}}\tilde{u}(j)$$

$$\geqq a_i^k+\sum_j p_{ij}^k\tilde{u}(j) \quad \text{for all } k$$

so that for

$$u=\tilde{u}, \quad a=\tilde{a}$$

$$u(i)+a=\underset{k}{\text{Max}}\left[a_i^k+\sum_j p_{ij}^k u(j)\right].$$

We have shown

Theorem: *Necessary and sufficient for $\hat{k}=\hat{k}(i)$ to be an optimal decision rule is the existence of a value function $u(i)$ and an associated average payoff a such that*

(10) $u(i)+a=\underset{k}{\text{Max}}\left[a_i^k+\sum_j p_{ij}^k u(j)\right].$

Policy iteration may be extended to the case of discounted returns. For any given decision rule $k = d(i)$ determine a unique value function

$$(11) \qquad v(i) = a_i^{d(i)} + \rho \sum_j p_{ij}^{d(i)} v(j).$$

Find an improved decision rule $\hat{d}(i)$ by maximizing the right hand side in terms of the provisional value function $v(j)$

$$(12) \qquad \begin{aligned} \text{Max}_k &\left[a_i^k + \rho \sum_j p_{ij}^k v(j) \right] \\ &= a_i^{\hat{d}(i)} + \rho \sum_j p_{ij}^{\hat{d}(i)} v(j) \\ &\geq v(i) \end{aligned}$$

and $>$ for at least some i.

A new value function $v(i)$ is then defined by

$$(13) \qquad \hat{v}(i) = a_i^{\hat{d}(i)} + \rho \sum_j p_{ij}^{\hat{d}(i)} \hat{v}(j).$$

Subtracting (12) from (13)

$$(14) \qquad [\hat{v}(i) - v(i)] \geq \rho \sum_j p_{ij}^{\hat{d}(i)} [\hat{v}(j) - v(j)]$$

$$> \text{ for some } i.$$

Consider the smallest $\hat{v}(i) - v(i)$, and suppose it is negative. Suppose the minimum is taken on for $i = i_0$. Now

$$\begin{aligned} \hat{v}(i_0) - v(i_0) &\geq \rho \sum_j p_{ij}^{\hat{d}(i)} [\hat{v}(j) - v(j)] \\ &\geq \rho \sum_j p_{ij}^{\hat{d}(i)} [\hat{v}(i_0) - v(i_0)] \\ &= \rho [\hat{v}(i_0) - v(i_0)]. \end{aligned}$$

Thus

$$[\hat{v}(i_0) - v(i_0)](1 - \rho) \geq 0.$$

Therefore,

$$(15) \qquad \hat{v}(i) \geq v(i) \qquad \text{for all } i \text{ and}$$

$$> \qquad \text{for some } i \text{ in view of (14)}.$$

Only a finite number of different decision rules $d(i)$, and hence, of equation systems (11) exist. Because of (15) no equation system may be repeated. But the algorithm can be continued as long as (12) applies to any i. Therefore, eventually

$$(16) \qquad a_i^k + \rho \sum_j p_{ij}^k v(j) \leq v(i) \qquad \text{for all } i.$$

The equality sign is taken on by choosing $k=\tilde{d}(i)$, where $\tilde{d}(i)$ is the last and optimal decision rule. Then

$$v(i)=\operatorname*{Max}_{k}\left[a_i^k+\rho\sum_j p_{ij}^k v(j)\right].$$

§ 14. Stability Properties

For any given decision rule the value function

$$v_n(i)$$

is continuous with respect to ρ. Suppose that the optimal decision rule for the undiscounted problem is unique. Then the value function under the second best rule differs by a finite amount. From the continuity of v in ρ it follows that a number ρ_0 exists such that the optimal policy is identical for all ρ in $\rho_0\leq\rho\leq1$.

From (11.2) it is seen that for $\rho<1$

$$|v(i)-v_n(i)|<\varepsilon \quad \text{for all } i \text{ and all } n>N(\rho).$$

It follows that there exists a $\hat{N}$ such that for given $\rho<1$, the optimal policy is identical for all decision chains of length $n>\hat{N}$. In the undiscounted problem equation (11.6) shows that

$$|u(i)-u_n(i)|<\varepsilon \qquad \text{for } n>\hat{N}.$$

Therefore the optimal policy is again the same for all $n>\hat{N}$.

Define now

(1) $$u_n(i,\rho)=v_n(i,\rho) - \frac{a\,(1-\rho^n)}{1-\rho},$$

for a given decision rule.

$$v_n(i,\rho)=\sum_{t=0}^{n} \sum_j p_{ij}^{(t)} \rho^t a_j$$

$$u_n(i,\rho)=\sum_{t=0}^{n} \sum_j p_{ij}^{(t)} \rho^t(a_j-a)$$

$$=\sum_{t=0}^{n} \sum_{r=1}^{m} \sum_j q_j^{(r)} \lambda_r^t \rho^t(a_j-a) \quad \text{using (10.6).}$$

Since

$$\sum_{j=1}^{m} q_j^{(1)} a_j = \sum_{j=1}^{m} \pi_j a_j = a$$

the term $q_j^{(1)}$ is cancelled out

$$u_n(i,\rho) = \sum_{t=0}^{n} \sum_{r=2}^{m} q_j^{(r)} \lambda_r^t \rho^t.$$

Since all $|\lambda_r| < 1$ it is seen that

(2) $\lim_{n\to\infty} u_n(i,\rho)$

exists uniformly in ρ for all $0 \le \rho \le 1$.
Therefore, the order of the limit processes may be interchanged

(3) $\lim_{\rho\to 1} \lim_{n\to\infty} u_n(i,\rho) = \lim_{n\to\infty} \lim_{\rho\to 1} u_n(i,\rho)$

$$= u(i).$$

It follows that there exist constants N_0, $\rho_0 < 1$ such that the infinite horizon decision problem without discounting has the same optimal policy as all decision problems with $n > N_0$ and $\rho > \rho_0$. The theoretical significance of this result is that in the infinite horizon undiscounted case the optimal policy is always stationary. This follows since that policy is identical with a stationary policy in the discounted case—based on a stationary value function. The practical importance of this result is that decision problems with long horizon and high discount factors may be investigated by solving the simpler infinite horizon undiscounted decision problem.

Approximation errors. Assume as usual that all $a_i^k > 0$ so that $v_n(i)$ is monotone increasing in n.
Compare

(4) $v(i) = \text{Max}_k \left[a_i^k + \rho \sum_j p_{ij}^k v(j) \right]$

$$= a_i^{\hat{k}} + \rho \sum_j p_{ij}^{\hat{k}} v(j),$$

(5) $v_n(i) = \text{Max}_k \left[a_i^k + \rho \sum_j p_{ij}^k v_{n-1}(j) \right]$

$$\ge a_i^{\hat{k}} + \rho \sum_j p_{ij}^{\hat{k}} v_{n-1}(j).$$

Taking differences

(6) $$v(i)-v_n(i)\leqq\rho\sum_j p_{ij}^{\hat{k}}[v(j)-v_{n-1}(j)].$$

Let ε_n denote the largest deviation of v_n from v

$$\varepsilon_n=\underset{i}{\mathrm{Max}}\,[v(i)-v_n(i)].$$

From (6) it follows that

(7) $$\varepsilon_n\leqq\rho\,\varepsilon_{n-1}\,.$$

The maximum error decreases at least at the rate of the discount factor. The percentage improvement is at least of the order of the interest rate. Practical experience shows that this estimate of the rate of improvement is quite close so that value iteration tends to be rather slow.

To estimate ε_n consider

$$\delta_n=\underset{i}{\mathrm{Max}}\,[v_n(i)-v_{n-1}(i)].$$

From

$$\begin{aligned}v_n&=\underset{k}{\mathrm{Max}}\left[a_i^k+\rho\sum_j p_{ij}^k v_{n-1}(j)\right]\\&=a_i^{\hat{k}}+\rho\sum_j p_{ij}^{\hat{k}}v_{n-1}(j),\quad\text{say.}\end{aligned}$$

Now $$v_{n-1}=\underset{k}{\mathrm{Max}}\left[a_i^k+\rho\sum_j p_{ij}^k v_{n-2}(j)\right]\geqq a_i^{\hat{k}}+\rho\sum_j p_{ij}^{\hat{k}}v_{n-2}(j)$$

by definition of a maximizer.
Subtracting

$$v_n-v_{n-1}\leqq\rho\sum_j p_{ij}^{\hat{k}}[v_{n-1}(j)-v_{n-2}(j)]$$

and so

$$\delta_n\leqq\rho\,\delta_{n-1}\,.$$

Now

$$\begin{aligned}\varepsilon_n&=\delta_{n+1}+\delta_{n+2}+\cdots\\&\leqq\delta_n\,[\rho+\rho^2+\cdots]\end{aligned}$$

Thus

(8) $$\varepsilon_n\leqq\delta_n\frac{\rho}{1-\rho}$$

since δ_n can be observed (while ε_n cannot).

This estimate is useful in controlling the degree of approximation achieved in value iteration.

§ 15. Solution by Linear Programming

Consider the undiscounted problem with infinite horizon. Let x_i^k denote the joint probability of being in state i and applying action k. The expected payoff is then

(1) $$\sum_{i,k} x_i^k p_{ij}^k (b_{ij}^k + b_i^k) = \sum_{i,k} x_i^k a_i^k = a.$$

Note that

(2) $$x_i^k \geqq 0$$

and

(3) $$\sum_{i,k} x_{ik} = 1.$$

Also

$$\sum_k x_i^k = \pi_i$$

is the state probability of state i.

Therefore the stationary state transition law must hold

(4) $$\sum_{i,k} x_i^k p_{ij}^k = \sum_k x_j^k.$$

Maximizing (1) subject to (2) $\cdots$ (4) is a Linear Program. Denote the dual variables associated with (3) and (4) as μ and λ_j respectively. The efficiency conditions [BECKMANN (1959)] state that the bi-linear function

$$\sum_{i,k} x_i^k a_i^k + \sum_i \lambda_i \left[-\sum_k x_i^k + \sum_{j,k} x_j^k p_{ji}^k \right] + \mu \left(1 - \sum_{i,k} x_i^k \right)$$

has a maximum with respect to x_i^k and a minimum with respect to λ_i and μ. A necessary condition for this is that

$$x_i^k \begin{Bmatrix} \geqq \\ = \end{Bmatrix} 0 \quad \text{according as} \quad a_i^k - \lambda_i + \sum_j \lambda_j p_{ij}^k - \mu \begin{Bmatrix} = \\ < \end{Bmatrix} 0.$$

It follows that

(5) $$\lambda_i + \mu \geqq a_i^k + \sum_j p_{ij}^k \lambda_j,$$

"$=$" only if $x_i^k > 0$.

Since each state is reached with positive probability we must have

$$\sum_k x_i^k > 0$$

for every i. Hence for each i there exists a k such that

$$x_i^k > 0$$

and the "$=$" holds in (5). We conclude

(6) $$\lambda_i + \mu = \underset{k}{\text{Max}}\left[a_i^k + \sum_j p_{ij}^k \lambda_j\right].$$

The optimal policy is then to make that x_i^k positive, i. e. to choose that action k in state i for which the principle of optimality is satisfied. Again the dual variables (efficiency prices) λ_i correspond to the value function. The dual variable μ represents the stationary average payoff.

We conclude this section with some remarks on non-ergodic problems.

Cyclic Markov chains. This case has been treated in effect in PART ONE. Under certainty and repetition of states the optimal policy is in fact always cyclical.

Transient states and all recurrent states communicating. The analysis of ergodic chains can be extended. Limiting state probabilities are unique but some may be zero. In policy iteration improvement is restricted to recurrent states.

Non-communicating recurrent states and transient states. The decision problem is ergodic for each communicating set of recurrent states and may be solved separately for each set. Identification of the several communicating sets may be troublesome.

For chains starting in transient states the stationary state probability depends on the initial transient state. There exist multiple solutions of the equations for the value function and of the principle of optimality. These are difficulties that make policy iteration less effective. For details the reader is referred to HOWARD [HOWARD (1960)].

In every case, however, value iteration will succeed.

§ 16. Machine Care

As the simplest example of a stochastic decision process consider a machine which may be in two states

> 0   failed
>
> 1   operating.

In each state two actions are possible

Table 1

		action	
		0	1
state	0	normal repair	express repair
	1	no maintenance	preventive maintenance

The transition probabilities p_{ij}^0 under actions "0" are assumed to be

Table 2

		j	
i		$\frac{1}{2}$	$\frac{1}{2}$
		$\frac{1}{4}$	$\frac{3}{4}$

and the payoffs a_{ij}^0 are assumed to be

Table 3

		j	
i		-8	-5
		7	10

Under action "1" we have transition probabilities p_{ij}^1

Table 4

		j	
i		$\frac{1}{3}$	$\frac{2}{3}$
		$\frac{1}{10}$	$\frac{9}{10}$

and payoffs a_{ij}^1

Table 5

	j	
i	-15	-12
	5	8

The expected one-period returns are

$$a_0^0 = \sum_j p_{0j} a_{0j} = \tfrac{1}{2}(-8) + \tfrac{1}{2}(-5) = -6.5$$

etc.

The resulting a_i^k are summarized in the following table

Table 6

	k	
i	-6.5	-13
	9.25	7.7

Value iteration—no discounting. The general formula is

$$(1) \qquad v_n(i) = \underset{k=0,1}{\text{Max}} \left[a_i^k + \sum_{j=0}^{1} p_{ij} v_{n-1}(i) \right] \qquad i=0,1.$$

Notice that we have not entered a constant term $\bar{v}$ so that the value function should increase by that amount from period to period.

Now let $v_0(i) = 0$. One has

$$v_1(0) = \text{Max}[a_0^0 + 0, a_0^1 + 0]$$
$$= \text{Max}[-6.5, -13] = -6.5,$$
$$v_1(1) = \text{Max}[9.25, 7.7] = 9.25,$$
$$v_2(0) = \text{Max}[-6.5 + \tfrac{1}{2}(-6.5) + \tfrac{1}{2}9.25, -13 + \tfrac{1}{3}(-6.5) + \tfrac{2}{3}9.25]$$
$$= \text{Max}[-5.13, -8.97] = -5.13,$$
$$v_2(1) = \text{Max}[9.25 + \tfrac{1}{4}(-6.5) + \tfrac{3}{4}9.25, 7.7 + \tfrac{1}{10}(-6.5) + \tfrac{9}{10}9.25]$$
$$= \text{Max}[14.56, 15.38] = 15.38,$$

and so on. Calculations for the first five periods are summarized in the following table

Table 7

n	0	1	2	3	4	5
$v_n(0)$	0	-6.5	-5.13	-1.38	3.33	8.42
$v_n(1)$	0	9.25	15.38	21.03	26.49	31.87

With increasing n the differences $v_n(i) - v_{n-1}(i)$ approach each other, for instance

$$v_5(0) - v_4(0) = 5.09; \quad v_5(1) - v_4(1) = 5.38.$$

The limiting value is $\bar{v} = 5.33$ (see below). The decisions underlying the value functions are

Table 8

n	1	2	3	4	5
$d_n(0)$	0	0	0	0	0
$d_n(1)$	0	1	1	1	1

In this particular example there is no further change in the decision rule after the second period. Normal repairs and preventive maintenance are the optimal policy.

Under this policy

$$\delta = \{d(0), d(1)\} = \{0, 1\}.$$

The transition probabilities are

Table 9

$$\begin{pmatrix} p_{00}^0 & p_{01}^0 \\ p_{10}^1 & p_{11}^1 \end{pmatrix} = \begin{pmatrix} \frac{1}{2} & \frac{1}{2} \\ \frac{1}{10} & \frac{9}{10} \end{pmatrix}.$$

The state probabilities are then determined by

(2) $$(\pi_0, \pi_1) \begin{pmatrix} \frac{1}{2} & \frac{1}{2} \\ \frac{1}{10} & \frac{9}{10} \end{pmatrix} = (\pi_0, \pi_1)$$

and

(3) $\pi_0 + \pi_1 = 1$.

The solution is

$$\pi_0 = \tfrac{1}{6} \qquad \pi_1 = \tfrac{5}{6}.$$

The average payoff per period is then

$$a = \pi_0 a_0^0 + \pi_1 a_1^1 = \tfrac{1}{6}(-6.5) + \tfrac{5}{6}(7.7) = 5.33.$$

Policy iteration. As a first policy assume

$$d_1(0) = 0, \qquad d_1(1) = 0.$$

The associated value function is then given by

(4) $v^1(i) + \bar{v}^1 = a_i^0 + \sum_j p_{ij}^0 v^1(j) \qquad i = 0, 1$.

In particular

$$v^1(0) + \bar{v}^1 = -6.5 + \tfrac{1}{2}v^1(0) + \tfrac{1}{2}v^1(1),$$
$$v^1(1) + \bar{v}^1 = \quad 9.25 + \tfrac{1}{4}v^1(0) + \tfrac{3}{4}v^1(1).$$

Setting $v^1(1) = 0$ and solving the first equation for $v^1(0)$

(5) $v^1(0) = -13 - 2\bar{v}^1$.

Substituting in the second equation

$$\bar{v}^1 = 9.25 - \frac{13}{4} - \frac{\bar{v}^1}{2}, \qquad \bar{v}^1 = 4$$

and from (6)

$$v^1(0) = -21.$$

In term of this first value function $v^1(i)$ a second decision rule is found by determining

$$\underset{k}{\text{Max}} \left[a_i^k + \sum_j p_{ij}^k v^1(j) \right]$$

in particular for state $i = 0$

$$\text{Max}\left[-6.5 + \tfrac{1}{2}(-21) + \tfrac{1}{2}0, \; -13 + \tfrac{1}{3}(-21) + \tfrac{2}{3}0 \right]$$
$$= \text{Max}\left[-17, \; -20 \right] = -17$$

and for state $i=1$

$$\text{Max}\left[9.25+\tfrac{1}{4}(-21)+\tfrac{3}{4}\cdot0,7.7+\tfrac{1}{10}(-21)+\tfrac{9}{10}\cdot0\right]$$
$$=\text{Max}[4,5.6]=5.6.$$

An improved decision rule $d_2(i)$ is therefore

$$d_2(0)=0, \quad d_2(1)=1.$$

The associated value function $v^2(i)$ is determined by

$$v^2(0)+\bar{v}^2=a_0^0+\sum_j p_{0j}^0 v^2(j),$$
$$v^2(1)+\bar{v}^2=a_1^1+\sum_j p_{1j}^1 v^2(j).$$

Specifically

$$v^2(0)+\bar{v}^2=-6.5+\tfrac{1}{2}v^2(0)+\tfrac{1}{2}v^2(1),$$
$$v^2(1)+\bar{v}^2=7.7+\tfrac{1}{10}v^2(0)+\tfrac{9}{10}v^2(1).$$

Setting $v^2(0)=0$ one obtains the solution

$$v^2(1)=23.67, \quad \bar{v}^2=5.33.$$

In terms of $v^2(i)$ we want to find

$$\text{Max}_k\left[a_i^k+\sum_j p_{ij}^k v^2(j)\right].$$

Specifically for $i=0$

$$\text{Max}\left[-6.5+\tfrac{1}{2}\cdot23.67,9.25+\tfrac{1}{10}23.67\right]=18.16$$

$$d_3(0)=0$$

and for $i=1$

$$\text{Max}\left[9.25+\tfrac{3}{4}23.67,7.7+\tfrac{9}{10}23.67\right]=29$$

$$d_3(1)=1.$$

From $d_3(i)=d_2(i)$ we conclude that this is the optimal policy.

Direct calculation of average payoff. In this simple example it is feasible to compare the four possible policies directly with respect to the average yield $\bar{v}$ under a policy δ. This yield is determined by

(6) $$\bar{v}^\delta=\pi_0^\delta a_0^{d(0)}+\pi_1^\delta a_1^{d(1)}$$

where the state probabilities $\pi_i^{d(i)}$ are determined by $i = 0, 1$

(7)
$$\pi_0^\delta = \pi_0^\delta p_{00}^{d(0)} + \pi_1^\delta p_{10}^{d(1)},$$
$$\pi_1^\delta = \pi_0^\delta p_{01}^{d(0)} + \pi_1^\delta p_{11}^{d(1)}.$$

The results are summarized in the following table

policy	$d(0)=0$ $d(1)=0$	$d(0)=1$ $d(1)=0$	$d(0)=0$ $d(1)=1$	$d(0)=1$ $d(1)=1$
average yield $\bar{v}$	4	$3\frac{2}{11}$	5.33	5

§ 17. Inventory Control

In ordinary maximum problems where all data are given with certainty DP is often convenient but never indispensable. Its advantage comes to the fore when the problem is one of facing risk sequentially, for then the idea of recursiveness which is embodied in the principle of optimality is appropriate. Thus, in situations under risk and uncertainty recursiveness is no longer a mathematical convenience but the only way to formalize decisions taken in sequence, each depending on information that is then available but is not known in advance. In subjecting such sequential decision problems to analysis DP has no rival.

The classical dynamic program under risk and uncertainty ist that of inventory control. A recursive equation involving a minimal operator was first formulated for optimal inventory policy by ARROW, HARRIS, and MARSCHAK [1951]. Subsequent work has been carried out by many authors (see references at end of chapter). We shall present the standard model in some detail with emphasis on the fundamentals but forego a discussion of the many variants and refinements which now fill a voluminous and growing literature. No apologies will be made for the various simplifying assumptions. Some of these are adequate approximations of reality and some others could be relaxed without changing the basic approach.

The inventory problem is concerned with the best policy of stocking a commodity—such as repair parts—against an imperfectly foreseen demand in an effort to strike an optimal balance between the various costs of storage, reordering and stock-outs. Specifically we consider the following model.

Assumptions. Let decisions be made at the beginning of certain periods of equal length, e. g., three months (the practice of many firms). At those times one observes the level y of the inventory and decides on a quantity x to be reordered. (If $x=0$ no order ist placed.) The relevant data are:

The demand for the commodity during the period ahead. While this is a random variable, whose value is of course unknown, we assume that we know its probability distribution. Let p_i, $i=0,1,\dots$ denote the probability that demand in the next period is for i units. We assume that demands in different periods are independent and are subject to identical probability distributions. Demand which is not filled in the given period is backlogged, and no demand is lost. This simplifies the principle of optimality. But the results are not different when all unfilled demand is lost except that there is in effect then a higher shortage penalty cost.

The cost of ordering is considered to be the sum of a fixed cost k for paperwork, transportation etc. and of a cost equal to the purchasing price, or per unit procurement cost c, times the quantity ordered x. Thus total ordering cost equals $k+cx$. It is assumed that delivery is made instantly or rather within such a short time relative to the length of a period that this may be neglected.

The cost of storage is taken to be proportional to the stock at the beginning of the period (but after the arrival of any orders) when it is largest, namely $hy\geqq0$.

The cost of shortage is measured at the end of the period when it is largest. Such a shortage cost reflects the loss of goodwill due to inconvenience or costs inflicted on customers and the risk of losing future custom. Somewhat inconsistently we also assume that the probability distribution of demand is not affected by past shortages. In practice it is not easy to measure shortage penalty costs properly. Fortunately no great precision is needed since the order of magnitude is all that matters in the final result. An alternative approach, sometimes preferred in practice, is to fix the probability or expected value of shortages at a level regarded as tolerable, say 1%. This can be shown to be mathematically equivalent to fixing the unit shortage cost at an appropriate level, but it is less convenient for the analysis.

We assume the shortage cost to be proportional to the size of the shortage. Now, if y is the stock level (when positive) or shortage level (when negative) then the shortage cost is

$$=\begin{cases} 0 & y\geqq0 \\ -gy & y<0. \end{cases}$$

Since demand is a random variable the actual cost of shortage is not known at the beginning of the period; what is known is the expected value conditional on the level of stock at the beginning of the period, after stock orders have been received. If that stock is y, then a shortage of size $i-y$ $(i>y)$ will arise with probability p_i. The expected value of the shortage is

(1)
$$\sum_{i=y+1} (i-y)p_i$$

and the expected shortage cost is g times this amount.

The sum of storage and expected shortage cost in a period depends only on the stock level (after receipt of orders) and will be denoted by f_y

(2)
$$f_y = hy + g \sum_{i=y+1} (i-y)p_i.$$

The cost of ordering will be rewritten as follows. Let x be the size of the order and

$$\delta(x) = \begin{cases} 1 & \text{if} \quad x>0 \\ 0 & \text{if} \quad x=0 \end{cases}$$

be the Kronecker function. Then the cost of ordering is

$$k\delta(x) + cx.$$

The total expected cost in a period with initial stock y and order size x is then given by

$$k\delta(x) + cx + f_y.$$

The principle of optimality. The problem of formulating an optimal ordering policy is this. Ordering now will influence the timing of future orders and, hence, the costs that will arise in future periods. How can this cost be properly anticipated? It is here that the fundamental idea of DP comes in: Introduce as an unknown but well defined function the total expected discounted cost of inventory control, conditional on the present state of the system and on the assumption that the best inventory policy will be followed. Set up a relationship defining this unknown function. When the function is determined the optimal inventory policy can be obtained by minimizing this function.

On what information or "state variable" do present and future inventory costs depend? Certainly relevant is the present intentory level. Whether this is all that is needed can be decided by experiment: it suffices if we succeed in formulating the principle of optimality without additional state variables. Denote this tentative minimal expected discounted cost by $\Phi(y)$. In the inventory literature it is also called the "loss function".

Now at the end of the period stock is $y+x-i$ with probability p_i. Here a negative value of $y+x-i$ denotes a shortage to be made up. Now the assumption enters that there is no interaction of demand between periods: demand in different periods is independent and subject to the same probability distribution p_i. This guarantees that the decision problem at the beginning of the next period is again the same as that which we are facing now, conditional on a new stock level of course. Therefore, the loss funtion Φ applies again. We do not know its value in advance since we do not know the stock level $y+x-i$. But we know its expected value, and since Φ itself was defined as an expected value this is all that will be needed, as we shall see. This expected value is

$$\sum_{i=0}^{\infty} \Phi(x+y-i)p_i$$

which must be discounted by the discount factor ρ.

Consider now the sum of costs in the next period plus the discounted value of the expected future cost starting after the next period

$$k\delta(x)+cx+f_{y+x}+\rho \sum_{i=0}^{\infty} \Phi(x+y-i)p_i\,.$$

This cost depends not only on the initial stock level y but also on the decision variable x. If we now choose x so as to minimize this sum of costs what do we have? By definition this must be the loss function $\Phi(y)$. And so

$$(3) \qquad \Phi(y)=\underset{x \geq 0}{\text{Min}}\left[k\delta(x)+cx+f_{x+y}+\rho \sum_{i=0}^{\infty} \Phi(x+y-i)p_i\right]$$

is our desired equation defining Φ, viz. the principle of optimality.

There is, however, this important caveat: we have not actually shown that y is the only relevant variable. For instance Φ might depend on the very first stock with which we ever started. There is, therefore, no guarantee that equation (3) is meaningful, i. e., has a well-defined solution. On the other hand if we can produce a solution based on state variable y then this is seen to be one possible optimal inventory policy, although it is left open whether a policy using additional information (such as the initial stock) might not be superior. In the case of the inventory problem, any other state variable can be shown to be irrelevant [ARROW, KARLIN, SCARF, Chapter 9].

Discussion: It is useful to separate out of Φ a cost term which reflects the unavoidable variable ordering cost. Let

$$(4) \qquad \Phi(y)=\varphi(y)+c\left(\frac{\rho\mu}{1-\rho}-y\right)$$

where $\mu = E[i]$ is the mean demand per period. Substituting in (3) we have the following equation in terms of φ

$$\varphi(y) + \frac{c\rho\mu}{1-\rho} - cy = \operatorname*{Min}_{x \geq 0} \left\{ k\delta(x) + cx + f_{x+y} \right.$$

$$\left. + \rho \sum_{i=0}^{\infty} \varphi(x+y-i)p_i + \frac{c\rho^2\mu}{1-\rho} - \rho c(x+y-\mu) \right\}$$

which upon ordering terms reduces to

$$\varphi(y) = \operatorname*{Min}_{x \geq 0} \left[k\delta(x) + c(1-\rho)(x+y) + f_{x+y} + \rho \sum_{i=0}^{\infty} \varphi(x+y-i)p_i \right].$$

If we now redefine the storage and shortage cost function

$$F_y = f_y + c(1-\rho)y$$

we have the simpler equation

(5) $$\varphi(y) = \operatorname*{Min}_{x \geq 0} \left[k\delta(x) + F_{x+y} + \rho \sum_{i=0}^{\infty} \varphi(x+y-i)p_i \right].$$

In order to solve this equation for φ the method of value iteration may be used. Let the 0th approximation be

$$\varphi_0(y) = F_y$$

and define the nth approximation in terms of the $(n-1)$st approximation as follows

(6) $$\varphi_n(y) = \operatorname*{Min}_{x \geq 0} \left[k\delta(x) + F_{x+y} + \rho \sum_{i=0}^{\infty} \varphi_{n-1}(x+y-i)p_i \right].$$

The sequence converges to the limit function $\varphi_\infty(y)$ which is the unique solution of (5) (see § 12).

Since the cost terms are bounded on any bounded interval convergence follows by observing that the sequence φ_n is bounded and monotone. The economic reason for this monotone increase is that with each iteration the planning horizon and, hence, the total cost taken into consideration increases. The value iteration method must succeed although convergence may be slow.

The s,S policy. We return to equation

(5) $$\varphi(y) = \operatorname*{Min}_{x \geq 0} \left[k\delta(x) + F_{x+y} + \rho \sum_{i=0}^{\infty} \varphi(x+y-i)p_i \right].$$

Suppose for the moment that there were no fixed ordering cost, $k=0$. Then the right hand side is a function of $x+y$ only. There exists an optimal $x+y=S$ such that

$$F_S + \rho \sum_{i=0}^{\infty} \varphi(S-i)p_i = \underset{x \geq 0}{\text{Min}} \left[F_{x+y} + \rho \sum_{i=0}^{\infty} \varphi(x+y-i)p_i \right].$$

This means that in every period an identical best starting inventory S is chosen

$$S = x+y.$$

The inventory policy is then

$$x = S-y$$

and we have merely to determine the parameter S in order to obtain the best policy. Let now a fixed ordering cost exist $k>0$. Then it is no longer true that the starting stock will always be brought to its optimal level S; but it remains true that whenever a positive order x is placed, it will be such as to restore the optimal level $x+y=S$. The question is now, for which values y it is best to order, i. e., to choose $x>0$ and for which not to order i.e. $x=0$. Clearly for $y \geq S$, $x=0$. Moreover by continuity for y near S

$$\varphi(y) < k + \varphi(S).$$

If ordering should ever be practicable then at some level $y=s$ sufficiently low one must have

$$\varphi(y) \geq k + \varphi(S).$$

This suggests that the optimal policy might be of the following form: There exists a critical stock level s such that no order is placed when actual stock exceeds s and order is placed when actual stock is less or equal to s; when an order is placed it is such as to make $x+y$ equal to the optimal starting stock S. Formally:

$$x = 0 \qquad \text{when} \qquad y > s,$$

$$x = S-y \qquad \text{when} \qquad y \leq s.$$

The optimality of this policy has been proved rigorously by H. SCARF [1959]. We shall not give this proof but turn to the implications of this so-called (s, S) policy. We observe first that as a result of knowing the structure of the optimal policy, we can write down $\varphi(y)$ explicitly as follows

$$(7) \quad \varphi(y) = \begin{cases} F_y + \rho \sum_{i=0}^{y-s} \varphi(y-i)p_i + \rho[k+\varphi(S)] \sum_{i=y-s+1}^{\infty} p_i & \text{if} \quad y > s \\ \\ F_s + k & \text{if} \quad y \leq s. \end{cases}$$

Write now

$$y = s + x,$$

$$\varphi(y) = u(y - s) = u(x),$$

$$P_x = \sum_{i=0}^{x} p_i.$$

In terms of u the inventory equation becomes

(8)
$$u(x) = F_x + [1 - P_x] \rho [k + u(D)] + \rho \sum_{i=0}^{x} u(x - i) p_i \qquad x > 0$$

$$u(x) = k + u(D) \qquad\qquad\qquad\qquad\qquad\qquad x \leq 0$$

where
$$D = S - s.$$

In order to solve (8) it is convenient to use the generating function technique. Define the generating function

(9)
$$\sum_{x=0}^{\infty} u(x) z^x = G(z).$$

The object is to convert (8) into a simpler equation in terms of the generating function. Multiplying (8) by z^x and summing we have

(10)
$$\sum_{x=0}^{\infty} u(x) z^x = \sum_{x=0}^{\infty} F_x z^x + \rho [k + u(D)] \sum_{x=0}^{\infty} [1 - P_x] z^x$$

$$+ \rho \sum_{x=0}^{\infty} \sum_{i=0}^{x} u(x - i) p_i z^x.$$

The last term may be transformed as follows

$$= \rho \sum_{x=0}^{\infty} \sum_{i=0}^{x} u(x - i) z^{x-i} p_i z^i.$$

Introducing $x - i = j$ as a new summation index

$$= \rho \sum_{j=0}^{\infty} \sum_{i=0}^{\infty} u(j) z^j p_i z^i$$

$$= \rho \sum_{j=0}^{\infty} u(j) z^j \sum_{i=0}^{\infty} p_i z^i.$$

Substituting in (10)

$$\sum_{x=0}^{\infty} u(x) z^x = \sum_{x=0}^{\infty} F_x z^x + \rho [k + u(D)] \sum_{x=0}^{\infty} [1 - P_x] z^x +$$

5*

$$+ \rho \sum_{i=0}^{\infty} p_i z^i \sum_{x=0}^{\infty} u(x) z^x$$

or

$$(11) \qquad \sum_{x=0}^{\infty} u(x) z^x = \frac{\sum_{x=0}^{\infty} F_x z^x + \rho [k + u(D)] \sum_{x=0}^{\infty} [1 - P_x] z^x}{1 - \rho \sum_{i=0}^{\infty} p_i z^i}.$$

Now

$$\frac{1}{1 - \rho \sum_{i=0}^{\infty} p_i z^i} = 1 + \rho \sum_{i=0}^{\infty} p_i z^i + \rho^2 \sum_{i=0}^{\infty} \sum_{j=0}^{i} p_j p_{i-j} z^i + \cdots$$

Note that

$$\sum_{j=0}^{i} p_j p_{i-j}$$

is the probability of i events in two trials, sometimes written $p_i^{(2)}$.
Generally

$$\frac{1}{1 - \rho \sum_{i=0}^{\infty} p_i z^i} = 1 + \rho \sum_{i=0}^{\infty} p_i z^i + \rho^2 \sum_{i=0}^{\infty} p_i^{(2)} z^i + \cdots + \rho^n \sum_{i=0}^{\infty} p_i^{(n)} z^i + \cdots.$$

For convenience set $\quad p_i = p_i^{(1)}$

$$p_i^{(0)} = \begin{cases} 1 & i = 0, \\ 0 & i > 0. \end{cases}$$

Then

$$\frac{1}{1 - \rho \sum_{i=0}^{\infty} p_i z^i} = \sum_{n=0}^{\infty} \rho^n \sum_{i=0}^{\infty} p_i^{(n)} z^i$$

$$= \sum_{i=0}^{\infty} \left(\sum_{n=0}^{\infty} \rho^n p_i^{(n)} \right) z^i$$

provided the usual requirements for the interchange of summations apply.
Returning to (11)

$$\sum_{x=0}^{\infty} u(x) z^x = \sum_{x=0}^{\infty} \sum_{i=0}^{x} F_{x-i} \sum_{n=0}^{\infty} \rho^n p_i^{(n)} z^x$$

$$+ \rho [k + u(D)] \sum_{x=0}^{\infty} \sum_{i=0}^{x} [1 - P_{x-i}] \sum_{n=0}^{\infty} \rho^n p_i^{(n)} z^x.$$

The power series on the right and left hand side must be termwise equal

$$u(x) = \sum_{i=0}^{\infty} F_{x-i} \sum_{n=0}^{\infty} \rho^n p_i^{(n)} + \rho [k + u(D)] \sum_{i=0}^{x} [1 - P_{x-i}] \sum_{n=0}^{\infty} \rho^n p_i^{(n)}.$$

In particular for $x = D$

$$u(D) = \sum_{i=0}^{D} \sum_{n=0}^{\infty} F_{D-i} \rho^n p_i^{(n)} + \rho[k + u(D)] \sum_{i=0}^{D} \sum_{n=0}^{\infty} [1 - P_{D-i}] \rho^n p_i^{(n)}.$$

Ordering terms we have

(12) $$u(D) = \frac{k \sum_{i=0}^{D} \sum_{n=0}^{\infty} \rho^{n+1} p_i^{(n)} (1 - P_{D-i}) + \sum_{i=0}^{D} \sum_{n=0}^{\infty} F_{D-i} \rho^n p_i^{(n)}}{1 - \sum_{i=0}^{D} \sum_{n=0}^{\infty} \rho^{n+1} p_i^{(n)} (1 - P_{D-i})},$$

$$= \varphi(S).$$

Since it is reasonable to assume that ordering takes place before shortages arise, $s \geq 0$. Therefore

$$\varphi(0) = \varphi(S) + k.$$

Substituting in (12) we have

$$\varphi(0) = \frac{k + \sum_{i=0}^{D} F_{D-i} \sum_{n=0}^{\infty} \rho^n p_i^{(n)}}{1 - \sum_{i=0}^{D} (1 - P_{D-i}) \sum_{n=0}^{\infty} \rho^{n+1} p_i^{(n)}}.$$

The probability terms in the denominator have the following interpretation

$$(1 - P_{D-i}) \sum_{n=0}^{\infty} p_i^{(n)}$$

is the probability that in n periods demand is for a total of i items and in the next period demand is for not less than $D - i$ items. If i is allowed to range from 0 to D this becomes the probability that $n + 1$ trials are required for total demand to exceed D.

Let $P_D^{(n)}$ denote the probability that in n trials a total demand of D or less occurs. Then

$$P_D^{(n)} = \sum_{i=0}^{D} p_i^{(n)},$$

$$P_D^{(n+1)} = \sum_{i=0}^{D} P_{D-i} p_i^{(n)}.$$

Thus

(13) $$\varphi(0) = \frac{k + \sum_{i=0}^{D} F_{D-i} \sum_{n=0}^{\infty} \rho^n p_i^{(n)}}{1 - \sum_{n=0}^{\infty} \rho^{n+1} [P_D^{(n)} - P_D^{(n+1)}]}.$$

The sum in the denominator may be transformed by observing that each term $P^{(n)}$ occurs twice: once with ρ^{n+1} and once with $-\rho^n$. Thus

$$1- \sum_{n=0}^{\infty} \rho^{n+1} \left[P_D^{(n)} - P_D^{(n+1)} \right]$$

$$= 1 - \rho P_D^{(0)} - \sum_{n=1}^{\infty} P_D^{(n)} (\rho^{n+1} - \rho^n)$$

$$= 1 - \rho P_D^{(0)} - (\rho-1) \sum_{n=1}^{\infty} \rho^n P_D^{(n)}.$$

Recalling the definition of $p_i^{(0)}$ one sees that $P_D^{(0)} = 1$

$$= (1-\rho) \sum_{n=0}^{\infty} \rho^n P_D^{(n)}.$$

Substituting in (13)

$$\varphi(0) = \frac{k + \sum\limits_{i=0}^{D} F_{D-i} \sum\limits_{n=0}^{\infty} \rho^n p_i^{(n)}}{(1-\rho) \sum\limits_{n=0}^{\infty} \rho^n P_D^{(n)}}.$$

In terms of f one has

$$(14) \qquad \varphi(0) = \frac{k + \sum\limits_{i=0}^{D} f_{s+D-i} \sum\limits_{n=0}^{\infty} \rho^n p_i^{(n)}}{(1-\rho) \sum\limits_{n=0}^{\infty} \rho^n P_D^{(n)}}.$$

We have thus found an expression for the present value of costs at the outset when stocks are zero, in terms of the policy parameters s and D. The problem is thus reduced to minimizing the right hand side of (14) with respect to s and D.

Case of no discounting. We now want to consider the probabilities π_y of the initial stocks being at various levels y and reconsider equation (14).

Let stock after the receipt of stock orders be at an initial level y

$$s \leq y \leq S.$$

First consider the case $y = S$. Then either the initial stock level in the preceding period was S and no demand arose, and this has probability

$$\pi_S p_0$$

or a demand did arise and was such that the stock level at the end of the period fell below s; the probability of this is

$$\sum_{y=s}^{s} \pi_y [1 - P_{y-s}].$$

Since these are independent possibilities π_S is the sum of the two probabilities

(15) $$\pi_S = \pi_S p_0 + \sum_{y=s}^{s} \pi_y [1 - P_{y-s}].$$

Consider next $y \neq S$.

Then in the preceding period, stock must have been at a level $i \geq y$ and demand must have been for exactly $i - y$ items. Thus

(16) $$\pi_y = \sum_{i=y}^{s} \pi_i p_{i-y} \qquad y \geq s.$$

Observe that the process is ergodic since every state y in $s \leq y \leq S$ can be reached from every other state in two periods or more.

Therefore, the solution of (16) is unique up to a factor of proportionality. It is

(17) $$\pi_y = \pi_S (1 - p_0) \sum_{n=1}^{\infty} p_{S-y}^{(n)} \qquad s \leq y < S$$

where $p_i^{(n)}$ is the compound probability of a demand for i items in n periods as defined above. The solution (17) may be verified by substitution in (16). It may also be shown that (15) is satisfied by (17) provided we fix the proportionality factor by the condition

(18) $$\sum_{i=s}^{s} \pi_i = 1.$$

Applying (18) to (17) and (15) one obtains

(19)
$$\pi_y = \pi_{S-i} = \frac{\displaystyle\sum_{n=0}^{\infty} p_i^{(n)}}{\displaystyle\sum_{n=0}^{\infty} P_D^{(n)}},$$

$$\pi_S = \frac{1}{1 - p_0} \frac{1}{\displaystyle\sum_{n=0}^{\infty} P_D^{(n)}}.$$

Returning to (14) we observe that

$$\lim_{\rho \to 1}(1-\rho)\varphi = \frac{k+\sum\limits_{n=0}^{\infty} p_i^{(n)} f_{S-i}}{\sum\limits_{n=0}^{\infty} P_D^{(n)}},$$

(20) $\varphi = k\,(1-p_0)\pi_S + \sum\limits_{y=s}^{S} \pi_y f_y = \bar\varphi$ (say).

It can now be seen that $\bar\varphi$ represents the average cost per period. With probability π_y the initial stock level after ordering is y and f_y represents the expected storage and shortage cost in this period conditional on this starting stock. With probability $(1-p_0)\pi_S$ an initial stock level of S occured which was not preceded by a period of no demand, and this is the probability that an ordering cost k was incurred.

In principle, the steady state probabilities of a system may always be calculated as exemplified here, once the optimal policy is known. In some cases this may even be the simpler approach and an alternative to DP.

Geometric demand distribution. To illustrate these results further and to indicate their use in obtaining a problem solution, consider the special case of a geometric distribution of demand

(21) $p_i = q\,p^i, \quad 0 < p < 1, \quad q = 1-p.$

The π_y are determined by substituting (21) in (17)

(22) $\pi_y = \pi_S(1-p_0) \sum\limits_{n=1}^{\infty} p_{S-y}^{(n)}, \quad y \neq S$

$$= \pi_S(1-q) \sum_{n=1}^{\infty} \binom{n+S-y-1}{S-y} q^n p^{S-y},$$

where we have used the fact that the n-times compounded geometric distribution is the so-called negative binomial distribution [FELLER, p. 218]

$$p_i^{(n)} = \binom{n+i-1}{i} q^n p^i.$$

(22) may be calculated directly. An easier way of determining the values of (22) is by means of "generating function technique". Observe that the

generating function of (21) is

$$g(z) = \frac{q}{1-pz}$$

and that of the n-times compounded distribution

$$g(z)^n = \left(\frac{q}{1-pz}\right)^n.$$

Now

$$\sum_{n=1}^{\infty} p_i^{(n)}$$

is the coefficient of z^i in the power series defined by

$$\sum_{n=1}^{\infty} g(z)^n = \frac{g(z)}{1-g(z)} = \frac{1}{1-g(z)} - 1$$

$$= \frac{1}{1-\left(\dfrac{q}{1-pz}\right)} - 1$$

$$= \frac{q}{p} \cdot \frac{1}{1-z}$$

$$= q/p \sum_{n=0}^{\infty} z^n.$$

Thus

$$\sum_{n=1}^{\infty} p_y^{(n)} = q/p = \text{constant for all} \quad y \neq S.$$

From (17)

$$\pi_y = \pi_S(1-p_0)\, q/p = q\,\pi_S.$$

Since

$$1 = \pi_S + \sum_{y=s}^{S} \pi_y = \pi_S + D\pi_y = \pi_S[1+Dq],$$

one has finally

$$\pi_S = \frac{1}{1+Dq}, \qquad \pi_y = \frac{q}{1+Dq} \qquad y \neq S.$$

Substituting these state probabilities in (20) we have

$$\bar{\varphi} = [(1-q)k+f_s]\,\frac{1}{1+Dq} + \frac{q}{1+Dq} \sum_{y=s}^{S-1} f_y.$$

Now

$$f_y = hy + g \sum_{i=y+1}^{\infty} (i-y)qp^i = hy + \frac{g p^{y+1}}{q}.$$

And so

$$\bar{\varphi} = \frac{pk + hS + g \dfrac{p^{S+1}}{q} + qh \sum_{y=s}^{S-1} y + qg \sum_{y=s}^{S-1} \dfrac{p^{y+1}}{q}}{1+Dq}$$

$$= \frac{pk + h \left[S + qDs + qD \left(\dfrac{D-1}{2} \right) \right] + g \left[\dfrac{p^{S+1}}{q} + \dfrac{p^{s+1} - p^{S+1}}{1-p} \right]}{1+Dq}$$

$$(23) \qquad = h \left(s + \frac{D-1}{2} \right) + \frac{kp + h \dfrac{D+1}{2} + g \dfrac{p^{s+1}}{q}}{1+qD}.$$

This is the loss function or average cost per period in terms of the policy parameters s and D. To obtain s and D we must minimize (23). This may be done by differencing: the optimal s is the smallest integer for which the first difference of (23) with respect to s is non-negative

$$\Delta_s \bar{\varphi} = h + \frac{g \dfrac{p^{s+1}}{q} (p-1)}{1+qD} \geqq 0$$

from which

$$(24) \qquad p^s \geqq \frac{h}{g} \frac{1+qD}{p} > p^{s+1}$$

or

$$(25) \qquad s = \frac{\log \left[\dfrac{h}{g} \dfrac{1+qD}{p} \right]}{\log p}.$$

Similarly D is the smallest integer for which the first difference with respect to D is non-negative. A straightforward calculation yields

$$(26) \qquad h \left[\frac{q^2 D}{2} (D+1) + qD + 1 \right] \geqq qpk + g p^{s+1}.$$

Substituting the approximation (24) for p^s one obtains after an easy calculation

$$(27) \qquad D(D+1) = \frac{2pk}{qh}.$$

Now $\dfrac{p}{q} = \mu$ is the expected value of demand per period. The solution (27) corresponds, therefore, closely to the so-called WILSON lot size formula

(28) $$D = \sqrt{\dfrac{sk\mu}{h}}$$

which has been known for a long time as the solution in the case where demand in each period equals μ *with certainty* [HADLEY and WHITIN].

It is a fortunate accident that the simple formula (27) remains valid for the geometric demand distribution. It can be shown that it gives a first approximation—which is often adequate—also for more complicated demand distributions, e.g., the members of the negative binomial family among which i has the highest variance/mean ratio.

In general, the state probabilities π_i which are the crucial part to the method of steady solutions are not as easily determined as in the case of the geometric distribution. In fact it can be shown that the π_y are constant —with the possible exception of π_s—only for the geometric distribution and for the binomial distribution

$$p_0 = q\,,$$
$$p_1 = 1 - q\,.$$

However if the average demand per period μ is small it can be shown that the π_i are approximately constant. For the Poisson distribution of demand one has for instance

$$p_i = \dfrac{\lambda^i}{i!}e^{-\lambda}$$

which is in fact

$$\pi_{s-i} = \sum_{n=1}^{\infty} \dfrac{(n\lambda)^i}{i!}$$

approximately constant for small values of i, provided λ is small.

On the assumption that π_y, $y \neq S$, are constant, formulae for the optimal s and D can be derived. Various approximations are known also in more general cases [cf., for instance, D. M. ROBERTS, 1959]. But the simplest way of obtaining solutions of prescribed accuracy for any given demand distribution is value or policy iteration using the original DP formulation. Extensive calculations have been made by WAGNER [1962].

From an economic point of view the investment in ordering cost plus an initial inventory are an example of the "point input—continuous output" type of investment whose output—the negative expected value of storage and shortage cost—falls from period to period. It is then opti-

mal to make a new point input after a while. What makes the economies
of the inventory problem interesting is the fact that the "wearing out" of
the initial investment is a random process which is being controlled
through periodic inspections so that the reinvestment decision can be
made conditional on the results of these inspections. The inventory
problem illustrates the pattern of a number of typical investment problems in which the decisions are sequential and based on recurrent
observation of the state of the system, e.g. replacement, repair, production
control.

§ 18. Uncertainty: Adaptive Programming

Let the probabilities p_{ij}^k be unknown. The decision maker must estimate them as he goes along. In order to act even initially he is supposed
to have some prior information as reflected in an a priori subjective
probability distribution. From observations made as the decision process
moves along the decision maker will calculate a posteriori probabilities
by means of Bayes' principle, thus revising his subjective probabilities in
the light of experience. This is the idea of adaptive programming. We will
consider it for the special case where the transition probabilities have the
form

$$(1) \qquad\qquad p_{ij}^k = p_{k(i)-j}.$$

An example is that of inventory control where i is the inventory level at
the beginning of a period, and k the inventory after stock replenishment
and $k-j$ is the demand. p_r is then the probability of a demand of size r. We
assume that we know the family to which the distribution p_r belongs but
not all its parameters. For instance let p_r be a number of the negative
binomial family

$$(2) \qquad\qquad p_r = \binom{h+r-1}{r}(1-p)^h p^r$$

and suppose that we know h but not p. This family includes the geometric
distribution $(h=1)$ and the Poisson distribution $(h=\infty)$.

Let the a priori distribution of the unknown parameter p be a Beta
distribution

$$(3) \qquad B(p,a,b) = \frac{(a+b-1)!}{(a-1)!(b-1)!}\, p^{a-1}(1-p)^{b-1}.$$

Suppose the values j_{t-i} $(i=1,\dots,m)$ of the random variable have been
observed. For convenience write

$$j_{t-i} = r_i.$$

The a posteriori distribution of p is

$$\mathrm{pr}(p|r_1,r_2,\ldots,r_m) = \frac{\prod_{i=1}^{m}\binom{h+r_i-1}{r_i}(1-p)^{mh}p^{r_1+r_2+\cdots+r_m}B(p,a,b)}{\int_0^1 \prod_{i=1}^{m}\binom{h+r_i-1}{r_i}(1-p)^{mh}B(p,a,b)\,dp}$$

$$= \frac{(1-p)^{mh+b-1}p^{r+a-1}}{\int_0^1 (1-p)^{mh+b-1}p^{r+a-1}\,dp}$$

where we have written $r=\sum_{i=1}^{m}r_i$

$$= \frac{\Gamma(r+a+mh+b)}{\Gamma(r+a)\Gamma(mh+b)}p^{r+a-1}(1-p)^{mh+b-1}$$

$$(4) \qquad = B(p,r+a,mh+b).$$

We have thus shown that for the distribution of the binomial family, with given exponent h, the a posteriori distribution of the unknown parameter p depends *only* on the sum of random variables r and on the number of trials m. Specifically, in the case of the negative binomial distribution the distribution of p is a Beta distribution with parameters $r+a$ and $mh+b$.

If the a priori distribution of p is the uniform distribution, then $a=b=1$, and we have the simpler formula

$$(5) \qquad \mathrm{pr}(p|r,m) = \frac{\Gamma(r+mh+2)}{\Gamma(r+1)\Gamma(mh+1)}p^{r}(1-p)^{mh}.$$

In terms of this a posteriori probability of p the probability distribution of j is now determined by the formula

$$\mathrm{pr}(j|r,m)=\int_0^1 \mathrm{pr}(j|p)\,\mathrm{pr}(p|r,m)\,dp$$

$$=\binom{h+j-1}{j}\int_0^1 (1-p)^{h}p^{j}\frac{\Gamma(r+a+mh+b)}{\Gamma(r+a)\Gamma(mh+b)}p^{r+a-1}(1-p)^{mh+b-1}\,dp$$

$$=\binom{h+j-1}{j}\frac{\Gamma(j+r+a)\Gamma(b+(m+1)h)}{\Gamma(j+h+r+a+mh+b)}\frac{\Gamma(r+a+mh+b)}{\Gamma(r+a)\Gamma(mh+b)},$$

$$(6) \quad \text{pr}(j|r,m) = \frac{\binom{h+j-1}{j-1}\binom{r+a+mh+b-1}{r+a-1}}{\binom{(m+1)h+r+j+a+b-1}{j+r+a-1}} \frac{b+mh}{b+(m+1)h}.$$

With a uniform a prior distribution $a=b=1$ the expression simplifies to

$$(7) \quad \text{pr}(j|r,m) = \frac{\binom{h+j-1}{j}\binom{mh+r+1}{r}}{\binom{(m+1)h+r+j+1}{j+r}} \frac{mh+1}{(m+1)h+1}$$

and for the geometric distribution $h=1$

$$(8) \quad \text{pr}(j|r,m) = \frac{\binom{m+r+1}{r}(m+1)}{\binom{m+2+r+j}{r+j}(m+2)} = \frac{(m+r+1)!(r+j)!(m+1)}{r!(m+r+j+2)!}.$$

The principle of optimality. Since all information about the probability of future events is contained in
1. the state variable i,
2. the horizon n,
3. the number of periods elapsed m, and
4. the sum of random variables r observed in the case of distributions
   of the binomial family

the optimal policy should depend only on these parameters. The same is
then true of the value function. Thus, whenever the number of periods of
observation and the sum of observed random variables contain all the
information about the unknown probability distribution, then value
function and decision rule will depend only on r and m, and on the current
state variable and on the horizon

$$v = v_{m,n}(i,r),$$

$$d = d_{m,n}(i,r).$$

The principle of optimality may now be stated as follows

$$(9) \quad v_{m,n}(i,r) = \text{Max}_k \left[a_i^k(r,m) + \rho \sum_j \text{pr}(j|r,m) v_{m+1,n-1}(k-j,r+j) \right]$$

with the boundary conditions

(10)
$$v_{m,0}(i,r)=0$$

$$v_{0,n}(i,0)=\underset{k}{\text{Max}}\left[a_i^k(0,0)+\rho\sum_j pr(j|0,0)v_{1,n-1}(k-j,j)\right]$$

where $pr(j|0,0)$ is the a priori probability distribution and

$$a_i^k(0,0)$$

the expected one period payoff based on the a priori distribution.

Consider the infinite horizon problem. The principle of optimality becomes

(11)
$$v_m(i,r)=\underset{k}{\text{Max}}\left[a_i^k(r,m)+\rho\sum_j p_{k-j}v_{m+1}(k-j,r+j)\right]$$

which involves three state variables: m, i, r. In the special case of the geometric demand distribution

$$v_m(i,r)=\underset{k}{\text{Max}}\left[a_i^k(r,m)+\rho\,\frac{(m+r+1)!(m+1)}{r!}\times\right.$$

$$\left.\times\sum_j\frac{(r+j)!}{(m+r+j+2)!}\,v_{m+1}(k-j,r+j)\right].$$

The decision rule is determined as the maximizer k of the right hand side of (11) for given i,r,m. Even in the case of an infinite horizon there are thus three state variables.

While in principle such a value function may be determined by policy iteration along the lines of § 13 the presence of three state variables makes this extremely difficult and cumbersome. The conclusion to be drawn from this general discussion of sequential decision making under uncertainty is that, in general, the calculation of optimal Bayesian policies runs into formidable difficulties. Only special cases of which the following is an example have been calculated.

Inventory control without fixed ordering cost. Consider an inventory problem without ordering cost [SCARF (1959)]. Proportional ordering cost is assumed to be absorbed as a rental charge included in the carrying cost for inventory. Let the probability distribution of demand be geometric

$$pr(j)=(1-p)p^j$$

with unknown p. If the a priori distribution of p is uniform then the a posteriori distribution for observed demands $r_1,\ldots,r_m$ is then given by (8).

Since ordering cost is zero the optimal policy is known to be the following: At the beginning of each period order (or dispose of) an amount such that the starting stock becomes optimal. It follows that the value function v should not depend on the initial stock i.

$$v_m(i,r) = v_m(r)$$

Hence the value determination equation has the simple form

$$(12) \qquad v_m(r) = \operatorname*{Min}_k \left[a^k(r,m) + \sum_j \operatorname{pr}(j|r,m) v_{m+1}(r+j) \right].$$

Since $a^k(r,m)$ is the only term on the right hand side containing k the optimal policy requires simply the minimization of $a^k(r,m)$. The value function $v_m(r)$ is then irrelevant.

Let h be the carrying cost, g the shortage penalty cost per unit of the commodity. Then,

$$(13) \qquad a^k = a^k(r,m) = h \sum_{j=0}^{k-1} (k-j)\operatorname{pr}(j|r,m) + g \sum_{j=k}^{\infty} (j-k)\operatorname{pr}(j|r,m)$$

$$= (h+g) \sum_{j=0}^{k-1} (k-j)\operatorname{pr}(j|r,m) + g(\mu - k)$$

where $\mu = \sum_{j=1}^{\infty} j\operatorname{pr}(j|r,m)$.

A straightforward calculation shows that

$$(14) \qquad \mu(r,m) = \frac{r+1}{m}.$$

Also

$$(1-p) \sum_{j=0}^{k-1} p^j (k-j) = \frac{-p}{1-p} + \frac{p^{k-1}}{1-p} + k.$$

Multiplying by $pr(j|r,m)$ and integrating we obtain

$$\sum_{j=0}^{k-1} \operatorname{pr}(j|r,m)(k-j) = -\frac{r+1}{m} + \frac{r+1}{m} \frac{\binom{m+r+1}{m}}{\binom{m+k+r+1}{m}} + k.$$

Thus finally

$$(15) \qquad a^k(r,m) = h\left(k - \frac{r+1}{m}\right) + (h+g)\frac{r+1}{m} \frac{\binom{m+r+1}{m}}{\binom{m+k+r+1}{m}}.$$

Taking first differences with respect to k the minimizing k is found to be the solution of

(16)
$$\frac{(r+k+1)(r+k+2)\dots(r+k+m+1)}{(r+1)(r+2)\dots(r+m+1)} = \frac{h+g}{h}.$$

For large m, r using STIRLING's approximation for $m!$ one obtains

(17)
$$\left(1+\frac{m+1}{r+k}\right)^k = \frac{h+g}{h}.$$

Observe that for large r, m

$$\frac{r+k}{m+1} \to \mu = \frac{p}{1-p}.$$

Substituting in (6) we obtain

(18)
$$\frac{h}{h+g} = p^k$$

in agreement with the newsboy formula for a geometric demand distribution with known parameter p. Formula (16) is, therefore, the Bayesian extension of the "newsboy rule" (18).

§ 19. Exponential Weighting

The distributions of the binomial type give rise to two sufficient statistics: the number of trials and the sum of the random variables. The resulting Dynamic Program involves three state variables in addition to the horizon.

Suppose however that the estimate of the parameter "mean demand" is revised not by BAYES' principle but on the basis of exponential weighting. Let μ denote the prior expectation of the random variable. In the light of the last observation r let this estimate be revised to

(1)
$$\mu' = \alpha r + (1-\alpha)\mu$$

where α is the weight assigned to the last observation. This is equivalent to an expectation function

$$\mu = \alpha[r_1 + (1-\alpha)r_2 + (1-\alpha)^2 r_3 + \cdots].$$

The application of this weighting function may be justified when the probability distribution of the random variable is itself subject to random shifts which have an equal chance α of occuring in any period. Note that it is not necessary to record the number of observations m. The principle of optimality therefore takes the form

(2) $$v_n(i,\mu) = \max_k \left[\sum_r p_r(\mu)a_{i,k-r}^k + \rho\sum_r p_r(\mu)v_{n-1}(k-r,\alpha r+(1-\alpha)\mu)\right].$$

For an infinite horizon this simplifies to a 2-state variable problem

$$(3) \quad v(i,\mu)=\underset{k}{\text{Max}}\left[\sum_r p_r(\mu)a^k_{i,k-r}+\rho\sum_r p_r(\mu)v(k-r,\alpha r+(1-\alpha)\mu)\right].$$

It is instructive to apply this procedure to the inventory problem with a geometric demand distribution and without ordering cost. The principle of optimality is

$$v(\mu)=\underset{k}{\text{Min}}\left[a^k(\mu)+\sum_r p_r(\mu)v(\alpha r+(1-\alpha)\mu)\right]$$

and here again minimization applies only to $a^k(\mu)$.

As before

$$a^k=(h+g)\sum_{r=0}^{k-1}(k-r)p_r+g(\mu-k)$$

with

$$\mu=\frac{p}{1-p}\quad\text{and}\quad p=\frac{\mu}{1+\mu},\quad p_r=(1-p)p^r.$$

Minimization yields the newsboy formula

$$(4) \qquad\qquad \frac{h}{h+g}=\left(\frac{\mu}{1+\mu}\right)^k$$

as an even simpler solution than (18.16).

For general demand distributions the result is similarly

$$(5) \qquad\qquad \text{pr}(\text{demand}\geq k|\mu)=\frac{h}{h+g}.$$

For the general problem (3) the calculation of the probability (5) is not that simple, particularly since μ is a continuous state variable. But value and policy iteration for problems in 2 state variables are not beyond our present computing capabilities, provided the range of i is not too large and the μ grid need not be too fine.

Quadratic criterion function. For special returns functions, the type of past information that may be considered can be quite general: when the returns are a quadratic function of the decision variable and when the resulting decision rule, which is then a linear function of the state variable, is linear also with respect to the past values of the random variable. This will be examined in § 26 below.

References and Selected Reading to Part Two

ARROW, K. J., T. HARRIS and J. MARSCHAK: Optimal Inventory Policy. Econometrica **19**, 3, 250–272 (1951). Erratum: **20**, 1, 133 (1952).

—, S. KARLIN and H. SCARF: Studies in the Mathematical Theory of Inventory and Production. Stanford: Stanford University Press 1958.

BECKMANN, M. J.: Lagerhaltung bei Unsicherheit. Unternehmensforschung **7**, 1, 9—26 (1963).

— Dynamic Programming and Inventory Control, Operations Research Quarterly **15**, 4, 389—400 (1964).

— On the Theory of Stochastic Control Processes. Bull. Soc. Royale Sciences Liège **33**, 520—529 (1964).

—, and D. HOCHSTÄDTER: Berechnung optimaler Lagerpolitiken. In: Jahrbücher für Nationalökonomie und Statistik 1968.

BELLMAN, R. E.: Adaptive Control Processes: A Guided Tour. Princeton: Princeton University Press 1961.

—, and S. E. DREYFUS: Applied Dynamic Programming. Ch. XI. Princeton: Princeton University Press 1962.

— Dynamic Programming and Markovian Decision Processes, with Particular Application to Baseball and Chess. In: Applied Combinatorial Mathematics, ed. E. F. Beckenbach. New York: Wiley 1964.

— Dynamic Programming and Inventory Control, Operations Research Quarterly **15**, 4, pp. 221—236. (1967).

BREIMAN, L.: Stopping-Rule Problems. In: Applied Combinatorial Mathematics, ed. E. F. Beckenbach. pp. 284—319. New York: Wiley 1964.

DE CANI, J. S.: A Dynamic Programming Algorithm for Embedded Markov Chains when the Planning Horizon is at Infinity. MS **10**, 4, 716—733 (1964).

DUBINS, L. E., and L. J. SAVAGE: How to gamble if you must. Inequalities for Stochastic Processes. London: McGraw Hill 1965.

FELLER, W.: An Introduction to Probability Theory and its Application. New York: Wiley, Vol. I, 1962, II; Vol. II, 1966, IV.

FLEMING, W. H.: Some Markovian Optimization Problems. JMM **12**, 1, 131—140 (1963).

GHELLINCK, G. DE: Applications de la Thèorie des Graphes: Matrices de Markov et Programmes Dynamiques. Cahiers du Centre d'Etudes de Recherche Operationnelle **3**, 1, 5—34 (1961).

HADLEY, G., and T. M. WHITIN: Analysis of Inventory Systems. Englewood Cliffs, New Jersey: Prentice Hall 1963.

HOWARD, R. A.: Dynamic Programming and Markov Processes. Cambridge, Mass.: MIT Press 1960.

IGLEHART, D. L.: Optimality of (s, S) Policies in the Infinite Horizon Dynamic Inventory Problem. MS **9**, 2, 259—267 (1963).

JEWELL, W. S.: The Properties of Recurrent-Event Processes. OR **8**, 4, 446—472 (1960).

KARLIN, S.: The Structure of Dynamic Programing Models. NRLQ **2**, 4, 285—294 (1955).

LINDLEY, D. V.: Dynamic Programming and Decision Theory. Applied Stat. (U. K.) **10**, 1, 39—51 (1961).

MAITRA, A.: A Note on Undiscounted Dynamic Programming. Ann. Math. Stat. **37**, 4, 1042—1044 (1966).

MANNE, A. S.: Linear Programming and Sequential Decisions. MS **6**, 3, 259—267 (1960).

MARSCHAK, J.: On Adaptive Programming, Statistical Control of Time Standards. MS **9**, 4, 517—541 (1963).

MURPHY, R. E.: Adaptive Processes in Economic Systems. New York–London: Academic Press 1965.

NEUMANN, J. VON, and O. MORGENSTERN: Theory of Games and Economic Behavior. 3rd ed. Princeton: Princeton University Press 1953.

ROBERTS, D. M.: Approximations to Optimal Policies in a Dynamic Inventory Model. In: ARROW, KARLIN, SCARF, Studies in Applied Probability and Management Science. Stanford 1962.

SASIENI, M. W.: A Markov Chain Process in Industrial Replacement. ORQ 7, 4, 148—155 (1956).

SCARF, H. E.: Bayes Solutions of the Statistical Inventory Problem. Ann. Math. Stat. 30, 2, 490—508 (1959).

— The Optimality of (S, s) Policies in the Dynamic Inventory. In: Mathematical Methods in the Social Sciences, K. J. ARROW, S. KARLIN, P. SUPPES, eds. Stanford: Stanford Univ. Press. 1960.

—, D. M. GILFORD, and M. W. SHELLY: Multistage Inventory Models and Techniques. Stanford: Stanford Univ. Press 1963.

WAGNER, H. M., and T. H. WHITIN: Dynamic Problems in the Theory of the Firm. NRLQ 5, 1, 53—74 (1958).

— Statistical Management of Inventory Systems. New York: Wiley 1962.

ZSCHOCKE, D.: Die Behandlung von Entscheidungsproblemen mit Hilfe des Dynamischen Programmierens. Unternehmensforschung 8, 3, 101—127 (1964).

PART THREE

Continuous Decision Variable

§ 20. An Allocation Problem

The simplest decision problem involving a continuous decision and state variable is the allocation of a limited resource to consumption (depletion) in various periods [BELLMAN 1957, 4–9; HOTELLING, 1931].

Let y be the total amount available. The decision is the optimal allocation among n periods (or possibly infinitely many). Consumption of an amount x in any period generates utility (social or private) $u(x)$. The utilities of different periods are made comparable through use of a utility discount factor ρ. The principle of optimality is

(1)
$$v_0(y) = 0 ,$$
$$v_n(y) = \underset{0 \leq x \leq y}{\text{Max}}[u(x) + \rho v_{n-1}(y-x)], \quad y = g(x, \xi) .$$

If u is continuous then as shall be shown, v_n is also continuous and since x is in a closed bounded set we are justified in using the Max operator rather than the supremum in (1).

In this particular case the problem is more easily solved directly, namely

(2)
$$\text{Max}\left[u(x_n) + \rho u(x_{n-1}) + \cdots + \rho^n u(x_1)\right]$$

subject to

(3)
$$x_n + x_{n-1} + \cdots + x_1 \leq y$$
$$x_k \geq 0 \quad k = 1, \ldots, n$$

which has the solution

(4)
$$\rho^{n-k} u'(x_k) = \lambda$$

where λ is a Lagrangean multiplier so determined that (3) holds (if "$<$" applies in (3) then $\lambda = 0$).

For the special utility function sometimes used in economic growth models

(5)
$$u(x) = x^a \quad 0 < a < 1$$

we have

$$\rho^{n-k} a x^{a-1} = \lambda$$

from which

(6)
$$x_k = \left(\frac{\lambda}{a}\rho^{k-n}\right)^{\frac{1}{a-1}}$$

and using (3) with $\lambda > 0$ we obtain

$$y = \sum x_k = \lambda^{-\frac{1}{1-a}} a^{\frac{1}{1-a}} \sum_{k=1}^{n} \left(\rho^{\frac{1}{1-a}}\right)^{n-k}$$

$$= \lambda^{-\frac{1}{1-a}} a^{\frac{1}{1-a}} \frac{1-\beta^n}{1-\beta}, \qquad \beta = \rho^{\frac{1}{1-a}},$$

$$\lambda = \left(\frac{a^{\frac{1}{1-a}}}{y}\frac{1-\beta^n}{1-\beta}\right)^{1-a} = a\, y^{a-1}\left(\frac{1-\beta^n}{1-\beta}\right)^{1-a},$$

so that finally

$$x_k = y^{a-1} = y\left(\frac{1-\beta}{1-\beta^n}\right)\beta^{n-k}.$$

Consumption in any period k is thus proportional to the total amount y and to

$$\beta^{n-k} = \rho^{\frac{n-k}{1-a}}$$

i. e., consumption shrinks by a constant factor β from period to period.

Of course, in the case of no discounting and identical utility functions the solution is to consume the same amount in each period.

Observe that consumption in the first period is

$$x_n = y\frac{1-\beta}{1-\beta^n}.$$

As the horizon extends to infinity, consumption in the first period decreases but converges to the finite amount

$$x = y(1-\beta),$$

a stationary policy.

The question arises whether this is true for all concave utility functions: Does there always exist a solution $x \geq 0$ to

(7)
$$v(y) = \max_{0 \leq x \leq y}[u(x) + \rho v(y-x)]$$

$$x = x(y)$$

and does $x = x(y)$ represent the optimal consumption plan in the infinite horizon allocation problem? To answer questions like these we must develop a theory of DP for continuous decision variables.

Consider also the possibility that the resource is growing, say that interest is accruing to an asset. This growth rate β need not be the reciprocal of the discount factor ρ since preference in allocation over time need not be determined by the growth of resources alone. It may be asked, however, whether the undiscounted allocation problem with growth

(8) $$v_n(y) = \underset{0 \le x_n \le \bar{y}}{\text{Max}} \left[u(x_n) + v_{n-1}(\beta(y - x)) \right]$$

differs significantly from the discounted one without growth (1)? For the special utility function $u = x^a$ they are in fact identical if $\dfrac{1}{\beta} = \rho$.

§ 21. General Theory

Consider the sequential decision problem involving a single state variable y and the single decision variable x from a closed bounded set A

(1) $$(x, y) \in A .$$

For example,

$$0 \le x \le c, \quad 0 < y < r < b \quad \text{and} \quad \alpha \le x \le \beta .$$

The choice of x may be further restricted by the horizon

(2) $$(x, y) \in S_n .$$

Let the one period payoff function be

$$f_n(x, y)$$

and assume it to be uniformly continuous and (hence) uniformly bounded on A

(3) $$f_n(x, y) < C \quad \text{for all } n.$$

Choice of a decision variable x in a state y produces a new variable

$$z = g_n(x, y) .$$

If for given y, $g_n(x, y)$ is monotone in x we may replace x by g as our decision variable. In the one period problem we now seek to

(4) $$\underset{\substack{x \\ (x, y) \in S_1}}{\text{Max}} f_1(x, y) = v_1(y) \quad \text{(say)}.$$

Since f is continuous on a closed bounded set, we are justified in using Max rather than Sup in (4).

The n period problem is now given by

(5)
$$\text{Max}_{\substack{x \\ (x,y)\in S_n}} [f_n(x,y)+\rho v_{n-1}(g_n(x,y))]=v_n(y)$$

or when g is monotone

(6)
$$\text{Max}_{\substack{x \\ (x,y)\in S_n}} [f_n(x,y)+\rho v_{n-1}(x)]=v_n(y).$$

Consider now a stationary problem in which f,g,S do not depend on n. Alternatively assume that S_n is non-decreasing, $S_n \supset S_{n-1}$, and that $f_n(y)$ is monotonously non-decreasing

$$f_{n+1}(y)\geq f_n(y) \quad \text{all } y.$$

Without restriction let $f(x,y)$ be non-negative. (If necessary add a constant without changing the decision problem as long as the horizon is fixed.) Now

$$|v_1(y)|=\text{Max } f(x,y)\leq c \leq \frac{c}{1-\rho}.$$

Suppose that $|v_n(y)| \leq \dfrac{c}{1-\rho}$ and consider

$$|v_{n+1}(y)|=\text{Max}[f(x,y)+\rho v_n(x)]$$

(7)
$$\leq c+\rho \frac{c}{1-\rho} = \frac{c}{1-\rho}.$$

Therefore, the $v_n(y)$ are uniformly bounded. Moreover, since $f(x,y)\geq 0$

$$v_1(y)\geq v_0(y)=0.$$

Assume that this is true for all v_k, $k=1,...,n$. Consider

$$v_{n+1}(y)= \text{Max}_{\substack{x \\ (x,y)\in S_n}} [f(x,y)+\rho v_n(x)]\geq \text{Max}_{\substack{x \\ (x,y)\in S_n}} [f(x,y)+\rho v_{n-1}(x)]$$

by induction hypothesis.
    Hence

(8)
$$v_{n+1}(y)\geq v_n(y).$$

Since the sequence of $v_n(y)$ is monotone non-decreasing and bounded it converges to a limit function $v(y)$.
    By construction this limit function satisfies the principle of optimality

(9)
$$v(y)= \text{Sup}_{\substack{x \\ (x,y)\in S}} [f(x,y)+\rho v(x)].$$

In fact the following more general proposition is true. Let

$$v_n(y) = \operatorname*{Sup}_{\substack{x \\ x \in S_n}} [f_n(x,y) + \rho\, v_{n-1}(g_n(x,y))]$$

exist for all n and suppose that for a sub-sequence n_k, $k=1,2,\ldots$

$$\lim_{k \to \infty} v_{n_k}(y) = v(y)$$

exists. Then $v(y)$ satisfies the principle of optimality

(10) $$v(y) = \operatorname*{Sup}_{x \in S} [f(x,y) + v(g(x,y))]$$

where S, f, g are the limits of the corresponding indexed functions. Now the supremum in (9) may be replaced by the maximum, since we have

Theorem 1: *Let $f(x,y)$ be non-negative and continuous in x and y on the closed bounded set S. Then the solutions $v_n(y)$ of (5) and $v(y)$ of (9) are continuous functions of y.*

Lemma 1: *Let $f(x,y)$ be continuous in x and y. Then*

$$\operatorname*{Max}_{\substack{x \\ (x,y) \in S}} f(x,y)$$

is a continuous function of y.

Proof: Let y_k $(k=1,\ldots)$ be an arbitrary convergent sequence $y_k \to \hat{y}$. Consider the sequence of x_k which individually maximize $f(x,y_k)$

$$f(x_k, y_k) = \operatorname*{Max}_{\substack{x \\ (x,y) \in S}} f(x,y_k).$$

In the closed bounded set S they contain a convergent subsequence $x_{k_r} \to \hat{x}$.

By construction

$$f(\hat{x}, \hat{y}) = \operatorname*{Max}_{\substack{x \\ (x,y) \in S}} f(x, \hat{y}) = v_1(\hat{y}).$$

By continuity of f $\lim_{y_k \to y} v_1(y_k) = \lim_{r \to \infty} f(x_{k_r}, y_{k_r}) = f(\hat{x}, \hat{y}) = v(\hat{y})$ and this

proves continuity of the maximum.

Lemma 1 shows in particular that $v_1(y)$ is continuous. Suppose now that $v_{n-1}(y)$ is continuous. In the same way as in the proof of the lemma one shows then continuity of $v_n(y)$.

To complete the proof of continuity of $v(y)$ we must show uniform convergence. Let x be the maximizer in $v_n(y)$ and consider

$$v_n(y) = f(\hat{x}, y) + \rho v_{n-1}(\hat{x}),$$

$$v_{n-1}(y) \geq f(\hat{x}, y) + \rho v_{n-2}(\hat{x}).$$

Subtract and observe that v_n is monotone non-decreasing

$$v_n(y) - v_{n-1}(y) \leq |v_n(y) - v_{n-1}(y)| \leq \rho |v_{n-1}(\hat{x}) - v_{n-2}(\hat{x})|.$$

Thus

$$|v_n(y) - v_{n-1}(y)| \leq \rho^{n-1} |v_1(\hat{x})| \leq \rho^{n-1} c < \varepsilon$$

for all y provided $n > N$. This shows that the sequence of continuous functions $v_n(y)$ converges uniformly on the closed bounded set S and therefore its limit is continuous. The principle of optimality may now be written

$$(11) \qquad\qquad v(y) = \underset{\substack{x \\ (x,y) \in S}}{\text{Max}} \; [f(x,y) + \rho v(x)].$$

Is every solution of (11) a value function for the infinite horizon decision problem? To show this we prove uniqueness. Suppose that $v(y)$ and $w(y)$ are two solutions of (11). Let x be the maximizer for v. Then

$$v(y) = f(\hat{x}, y) + \rho v(\hat{x})$$

and by definition of a maximizer

$$w(y) \geq f(\hat{x}, y) + \rho w(\hat{x}).$$

Subtracting

$$v(y) - w(y) \leq \rho [v(\hat{x}) - w(\hat{x})].$$

Repeat the process for $y = \hat{x}$ and write $x^{(n)}$ for the nth outcome

$$v(y) - w(y) \leq \rho^n [v(x^{(n)}) - v(x^{(n)})] < \varepsilon$$

for sufficiently large n since v, w are bounded. Thus

$$v(y) \leq w(y).$$

Exchanging their roles one proves in the same way

$$v(y) \geq w(y)$$

so that $v(y) \equiv w(y)$. We have proved

Theorem 2: *If $f(x, y)$ is continuous over the closed bounded set S, then the solution of the principle of optimality is unique and represents the value function of the infinite horizon decision problem.*

Theorem and proof are easily extended to non-stationary problems with uniformly convergent

$$f_n(x, y), \quad g_n(x, y) \quad \text{and} \quad S_n.$$

The assumption of boundedness is essential. That the theorem need not be true for unbounded functions f has been demonstrated by RADNER [(1967), p. 24]. Consider

(12)
$$v(y) = \operatorname*{Max}_{0 < x \leq \frac{y}{2}} \left[\log y + \tfrac{1}{2} v(x) \right].$$

The optimal solution is clearly $x = y/2$ and the value function

$$v(y) = 2 \log y - 2 \log 2.$$

However

$$\varphi(y) = v(y) - 1/y$$

is also a solution of the principle of optimality (12), and it does not generate the optimal policy.

The following properties of the value function may also be proved by induction.

Monotonicity. Suppose that for every x, $f(x, y)$ is a monotone increasing function of y and that $y_1 > y_0$

$$(x, y_0) \in S$$

implies

$$(x, y_1) \in S.$$

Then $v_n(y)$ and $v(y)$ are montone non-decreasing.

Proof: Since

$$f(x, y_1) > f(x, y_0)$$

and

$$(x, y_0) \in S \Rightarrow (x, y_1) \in S,$$

$$v_1(y_1) = \operatorname*{Max}_{\substack{x \\ (x, y_1) \in S}} f(x, y_1) > \operatorname*{Max}_{\substack{x \\ (x, y_0) \in S}} f(x, y_0) = v(y_0).$$

Now suppose that $v_{n-1}(y)$ is monotone increasing. For given x then so is

$$f(x, y) + \rho v_{n-1}(y)$$

and by the previous argument

(13)
$$\operatorname*{Max}_{x \in S_y} [f(x, y) + \rho v_{n-1}(y)]$$

is monotonically increasing.

Going to the limit one shows that $v(y)$ is monotone non-decreasing.

Homogeneity. Suppose $f(x,y)$ is homogeneous of degree h

$$f(\lambda x, \lambda y) = \lambda^h f(x,y)$$

and that $(x,y) \in S$ implies $(\lambda x, \lambda y) \in S$.

Then $v_n(y)$ and $v(y)$ are homogenous of degree h in y.

Proof:

$$v_1(\lambda y) = \underset{\substack{\lambda x \\ (\lambda x, \lambda y) \in S}}{\text{Max}} f(\lambda x, \lambda y),$$

$$= \lambda^h \underset{\substack{x \\ x, y \in S}}{\text{Max}} f(x,y) = \lambda^h v_1(y).$$

Let

$$v_{n-1}(\lambda y) = \lambda^h v_{n-1}(y),$$

$$v_n(\lambda y) = \underset{\substack{\lambda x \\ (\lambda x, \lambda y) \in S}}{\text{Max}} \left[f(\lambda x, \lambda y) + \rho v_{n-1}(\lambda y) \right]$$

$$= \underset{(x,y) \in S}{\text{Max}} \left[\lambda^h f(x,y) + \rho \lambda^h v_{n-1}(y) \right]$$

$$= \lambda^h v_n(y).$$

Going to the limit one shows that

(14) $$v(\lambda y) = \lambda^h v(y)$$

as asserted.

Positive homogeneity. Theorem and proof apply also when the sign of λ is restricted to be non-negative. The set S is then said to be a cone.

Theorem 3: Concavity. *Suppose that* $f(x,y)$ *is concave jointly in x and y and that S is convex. Then $v_n(y)$ and $v(y)$ are concave.*
To prove this we need the following

Lemma 2: *Let* $f(x,y)$ *be a jointly concave function of x and y and let S be convex. Then*

$$\underset{\substack{x \\ (x,y) \in S}}{\text{Max}} f(x,y)$$

is a concave function of y.

Proof: Concavity of $f(x,y)$ means

$$f(\lambda x + \mu x^1, \lambda y + \mu y^1) \geq \lambda f(x,y) + \mu f(x^1, y^1)$$

for all $\lambda, \mu \geq 0$, $\lambda + \mu = 1$, and any pair of points (x, y), (x^1, y^1).
Let

$$x \text{ be the maximizer of } f(\xi, y),$$
$$x^1 \text{ be the maximizer of } f(\xi, y^1).$$

Then

(15) $\qquad \lambda \underset{x}{\operatorname{Max}} f(x, y) + \mu \underset{x}{\operatorname{Max}} f(x^1, y^1) \leq f(\lambda x + \mu x^1, \lambda y + \mu y^1)$

$$\leq \underset{\xi}{\operatorname{Max}} f(\xi, \lambda y + \mu y^1)$$

and this proves concavity of

$$\underset{\substack{x \\ (x, y) \in S}}{\operatorname{Max}} f(x, y).$$

Proof of theorem 3 (Induction). Lemma 2 shows that $v_1(y)$ is concave. Let $v_{n-1}(y)$ be concave. Then

$$f(x, y) + \rho v_{n-1}(y)$$

is concave and by lemma 2 so is

(16) $\qquad \underset{\substack{x \\ (x, y) \in S}}{\operatorname{Max}} [f(x, y) + \rho v_{n-1}(y)] = v_n(y).$

Therefore $v_n(y)$ is concave for all n, and since this property is preserved in the limit $v(y)$ is concave as asserted.

Returning to the allocation problem, since it satisfies the assumptions of theorem 2, a solution v which does not vanish identically is obtained by successive substitution in (20.1), that is by value iteration. It is the unique solution of (20.1) and represents the value function of the optimal policy. If $u(x)$ is concave it has a well defined maximum. Suppose that

$$u(0) = 0$$

and

(17) $\qquad u(x) > 0 \quad \text{for} \quad x > 0.$

Then initial consumption must be positive, for

$$v(y) = u(0) + \rho v(y) \quad 0 \leq \rho < 1$$

would imply $v(y) \equiv 0$ in contradiction to

$$v(y) \geq v_1(y) = \underset{0 \leq x \leq y}{\operatorname{Max}} u(x) > 0$$

for $y > 0$.

Thus, a positive stationary consumption plan is assured for all concave utility functions satisfying (17).

§ 22. Linear Inhomogeneous Problems

We assume that the functions $f_n(x, y)$ are linear

(1) $$f_n(x, y) = b_{0n} + b_{1n} x + b_{2n} y$$

and that the constraint functions are linear

(2) $$a^n_{i1} x + a^n_{i2} y = c^n_i \qquad i = 1, \dots, m.$$

The general theory of linear problems will be given in connection with the case of several decision variables. Here we merely want to present some examples.

Warehousing [DANTZIG, p. 55, 56]. Suppose prices $p_0, p_1, \dots, p_n$ for successive periods are given for a commodity which may be stored in a warehouse of capacity 1.

Find the optimal storage policy.

Let y be the amount in storage, x the net addition (reduction if negative) to stock and consider the problem with horizon n

(3) $$v_0(y) = p_0 y,$$

(4) $$v_n(y) = \operatorname*{Max}_{\substack{x \\ 0 \leq x + y \leq 1}} \left[-p_n x + v_{n-1}(y + x) \right].$$

Write $x + y = z$. Then

$$v_n(y) = \operatorname*{Max}_{0 \leq z \leq 1} \left[p_n(y - z) + v_{n-1}(z) \right],$$

(5)
$$= p_n y + \operatorname*{Max}_{0 \leq z \leq 1} \left[-p_n z + v_{n-1}(z) \right]$$

shows that the state z to be attained next with $n - 1$ periods remaining is independent of the present state y, and depends only on future prices $p_{n-1}, p_{n-2}, \dots, p_0$. Therefore, the solution is never changed as the horizon is extended. We assert

$$v_n(y)$$

is a linear function of y

(6) $$v_n(y) = a_n + b_n y.$$

For $n = 0$ this is shown by (3).

Suppose (6) to be true for n. Then from (5)

$$v_{n+1}(y) = p_{n+1} y + \operatorname*{Max}_{0 \leq z \leq 1} \left[-p_{n+1} z + a_n + b_n z \right].$$

But the maximum of the linear function in brackets is taken on for $z=0$ or $z=1$ proving the assertion.

We have shown, incidentally, that the optimal policy can be chosen among those which involve only full and empty warehouse. The recursive algorithm (5) gives a very convenient method of solving this linear program.

Production smoothing, [S. JOHNSON (1957)]. Consider the short run problem of choosing between overtime production and storage to meet a non-stationary demand.

Let
- r_n demand, given
- y stock on hand, $y \geq 0$
- x production rate, $0 \leq x \leq 1$
- $h(y)$ storage cost
- $c(x)$ production cost.

Production cost is assumed to be piecewise linear and convex, storage cost to be linear. The output rate is restricted.

The object is to meet the given demand at minimum cost. The principle of optimality states

$$(7) \qquad v_0(y) = \begin{cases} 0 & y \geq r_0 \\ \infty & y < r_0 \end{cases}$$

$$(8) \quad v_n(y) = \underset{0 \leq x \leq 1}{\text{Min}} \left[c(x) + h(y+x-r_n) + v_{n-1}(y+x-r_n) \right] \quad \text{if} \quad y+1-r_n \geq 0$$

and $v_n(y) = \infty$ otherwise.

Feasibility requires obviously that

$$(9) \qquad \sum_{k=0}^{n} r_k \leq n .$$

If (9) holds we can drop the alternative $v_n = \infty$.

Notice that the solution does not involve r_{n+1}, r_{n+2} etc., and hence remains unchanged when the horizon is extended but y is kept the same. Of course, y may change as the horizon is extended.

The value function is now a piecewise linear function of y. The logic of the problem is not affected and the computation need not become more difficult when $c(x)$ and $h(y)$ are allowed to be non-linear convex functions. If c and/or h are non-convex the problem is difficult in ordinary mathematical programming, but remains a well defined Dynamic Program solvable by recursion.

Dam operation [LITTLE, 1955]. Let

w_n water inflow, given

y water stock, $0 \leq y \leq 1$

x water outflow, $0 \leq x \leq c$.

A utility function (concave, piecewise linear) $u(x)$ gives the value of water flow for electricity generation, irrigation etc. For feasibility it is sufficient that total flow in n periods never exceed nc. The principle of optimality gives the value of water in the reservoir

$$v_0(y) = 0,$$
(10)
$$v_n(y) = \max_{\substack{0 \leq x \leq c \\ 0 \leq y + w_n - x \leq 1}} [u_n(x) + v_{n-1}(y + w_n - x)].$$

Utility and water flow depend on the season and so on n. If there is strict periodicity in u_n and w_n then the problem with infinite horizon will have a periodic value function and, hence, a periodic solution with the same period.

§ 23. A Turnpike Theorem

As a problem in two variables consider the following version of the turnpike problem. Initially let stocks x_n, y_n of two commodities be given. Of these let the amounts

$$\xi_n \geq 0, \quad x_n - \xi_n \geq 0, \quad \eta_n \geq 0, \quad y_n - \eta_n \geq 0$$

be allocated to the production of outputs of the two commodities in the next period. Let the production functions be

(1)
$$x_{n-1} = f(\xi_n, \eta_n),$$
$$y_{n-1} = g(x_n - \xi_n, y_n - \eta_n).$$

These production functions are assumed to be homogeneous of degree one, and to be jointly concave, and increasing functions of the inputs. Moreover let both inputs be necessary

(2)
$$f(0, y) = f(x, 0) = 0,$$
$$g(0, y) = g(x, 0) = 0.$$

The value function is defined by means of a terminal payoff

(3)
$$v_0(x, y) = u(x, y)$$

where u is concave increasing and homogeneous of degree 1 and by the principle of optimality

(4)
$$v_n(x, y) = \max_{\substack{\xi, \eta \\ 0 \leq \xi \leq x \\ 0 \leq \eta \leq y}} v_{n-1}(f(\xi, \eta), g(x - \xi, y - \eta)).$$

Lemma:
$$v_n(x, y)$$

is a concave function of x *and* y.

Proof:
$$v_0(x, y)$$

is concave by assumption; assume

$$v_{n-1}(x, y)$$

to be concave.

Consider

$$\lambda v_n(x_1, y^1) + \mu v_n(x^2, y^2) = \lambda \rho \operatorname*{Max}_{\xi, \eta} v_{n-1}\big(f(\xi, \eta), g(x^1 - \xi, y^1 - \eta)\big)$$

$$+ \mu \rho \operatorname*{Max}_{\xi, \eta} v_{n-1}\big(f(\xi, \eta), g(x^2 - \xi, y^2 - \eta)\big)$$

$$= \lambda \rho v_{n-1}\big(f(\xi^1, \eta^1), g(x^1 - \xi^1, y^1 - \eta^1)\big)$$

$$+ \mu v_{n-1}\big(f(\xi^2, \eta^2), g(x^2 - \xi^2, y^2 - \eta^2)\big)$$

$$= \rho v_{n-1}\big(\lambda f(\xi^1, \eta^1) + \mu f(\xi^2, \eta^2), \lambda g(x^1 - \xi^1, y^1 - \eta^1)$$

$$+ \mu g(x^2 - \xi^2, y^2 - \eta^2)\big)$$

by induction hypothesis

$$\leq \rho v_{n-1}\big(f(\lambda \xi^1 + \mu \xi^2, \lambda \eta^1 + \mu \eta^2), g\big(\lambda(x^1 - \xi^1) + \mu(x^2 - \xi^2),$$

$$\lambda(y^1 - \eta^1) + \mu(y^2 - \eta^2)\big)\big)$$

by concavity of the production functions f, g

$$\leq \operatorname*{Max}_{\xi, \eta} \rho v_{n-1}\big(f(\xi, \eta), g(\lambda x^1 + \mu x^2 - \xi, \lambda y^1 + \mu y^2 - \eta)\big)$$

by definition of a maximum

$$= \rho v_n(\lambda x^1 + \mu x^2, \lambda y^1 + \mu y^2)$$

by (4).

Going to the limit shows that $v(x, y)$ is also concave. Let us make the additional assumption that the maximizing outputs

$$\hat{x} = f(\xi, \eta), \quad \hat{y} = g(x - \xi, y - \eta)$$

are continuous functions of (x, y). Consider the closed bounded set

$$x + y = 1, \quad 0 \leq x, y.$$

Then

$$\left(\frac{\hat{x}}{\hat{x} + \hat{y}}, \quad \frac{\hat{y}}{\hat{x} + \hat{y}} \right)$$

are also continuous functions of (x, y) and the mapping is "into". By the Brouwer fixed point theorem there exists then a fixed point, i.e. an x^*, y^* such that

7   Beckmann, Dynamic Programming

(5)
$$\frac{\hat{x}^*}{\hat{x}^*+\hat{y}^*} = x^* \qquad \text{and that}$$

$$\frac{\hat{y}^*}{\hat{x}^*+\hat{y}^*} = y^*$$

i.e.
$$\hat{x}^* = \lambda x^*,$$
$$\hat{y}^* = \lambda y^*.$$

Now
$$x^* = 0, \quad y^* = 1$$

is impossible since then

$$\hat{x}^* = \hat{y}^* = 0$$

by assumption (4) and similarly $x = 1, y = 0$ is impossible. Therefore

$$0 < x^*, \quad y^* < 1.$$

Now
$$v_n(x^*, y^*) = \rho v_{n-1}(\hat{x}^*, \hat{y}^*) = \rho v_{n-1}(\lambda x^*, \lambda y^*) = \rho \lambda v_{n-1}(x^*, y^*).$$

Introducing $\lambda^{-1} = \rho$ as a discount factor in (4) makes

$$v_{n-1}(x^*, y^*) = v_n(x^*, y^*) = v(x^*, y^*) = v_1(x^*, y^*) = u(x^*, y^*).$$

The limit function $v(x, y)$ has now the following properties:
It is concave and, hence, concave on the line segment

$$x + y = 1, \quad 0 \leqq x, y.$$

It vanishes at the endpoints

$$v(0,1) = v(1,0) = 0$$

and it has a non-zero value at an interior point $(x^*, 1 - x^*)$

$$v(x^*, 1 - x^*) \neq 0.$$

It follows that the concave function v must lie above the triangle spanned by the ordinate of x^* and the endpoints (Fig. 1).

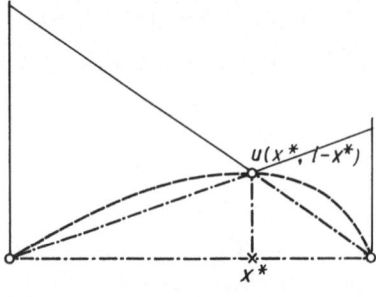

Fig. 1

Moreover, it must be bounded from above by concavity. Hence

$$v(x, 1-x) \neq 0$$

and by virtue of homogeneity

$$v(x, y) \neq 0 \quad \text{all} \quad x, y > 0.$$

This means that asymptotically the growth rate λ which applies from period to period is the same for all initial conditions and that it equals the growth rate along the paths which preserve the factor ratio—the von Neumann path. We must conclude that all growth paths run close to the von Neumann path. This is the essence of the celebrated turnpike theorem. The proof given here is readily extended to more than two factors.

§ 24. Sequential Programming

The theory developed in §21 can be extended in a straightforward manner to the case of several continuous decision variables. The most interesting case that arises is when $f(x, y)$ is linear or concave and the set S is a polyhedron. We want to show that these Dynamic Programs are Linear Programs. In the following let x, y be vectors, let A_n, B_n, G_n be matrices and h_n, f_n, c_n be coefficient vectors. Assume $v_0(y) = 0$

$$(1) \qquad v_n(y) = h_n y + \underset{x \geq 0}{\text{Max}} \{ f_n x + v_{n-1}(y - G_n x) \},$$
$$A_n x + B_n y \leq c_n.$$

Through recursion it is seen that (1) is a linear program sequentially compounded. In particular it is easily shown by induction that $v_n(y)$ is a piecewise linear concave function of y.

Of particular interest are problems in which only one or a small number of constraints are joint. Let y (scalar) denote the amount of a single resource to be used by different firms k $(k = 1, ..., n)$ which are subject to specific constraints

$$A_k x_k \leq c_k.$$

Then (1) assumes the special form

$$(2) \qquad v_n(y) = \underset{\substack{x \geq 0 \\ A_n x \leq c_n}}{\text{Max}} [b_n x + v_{n-1}(y - g_n x)].$$

Here the vectors g_n represent the coefficients for the inputs of joint resources (labor) into the activities of the different firms. The principle of optimality (2) is a method of decomposing a complex linear program into smaller programs tied together by some joint resources.

7*

The principle of optimality constitutes an alternative to the decomposition principle of DANTZIG/WOLFE [DANTZIG (1963), ch. 23]. In particular

$$v_{n-1}(y)$$

measures the value of the alternative uses of the resource, it is the opportunity cost function. Decompositon by the principle of optimality is a practical approach only when the number of joint resources (state variables) is small, say one or two. To an infinite horizon would correspond an unlimited number of firms. Notice also that this decomposition method does not really require linearity [cf. also NEMHAUSER (1964)].

§ 25. Risk

As before in § 12 we let payoff and transition to a new state depend not only on state and decision but also on a random variable u of known distribution $dP(u)$. The expected one-period payoff is then a function of the state variable and decision only. For convenience we may identify the new state with the random variable. Its distribution is then conditional on x and y.

$$P(u|x,y)$$

represents a general Markov process.

The principle of optimality is

(1)
$$v_n(y) = \underset{\substack{x \\ (x,y) \in S}}{\text{Max}} \left[f(x,y) + \rho \int v_{n-1}(u) dP(u|x,y) \right],$$

$$v_0(y) = 0.$$

However, when the variable for the new state may be written as a linear function of the old state y, of the decision variable x and of the random variable u, it is more convenient to set

(2)
$$v_n(y) = \underset{x, y \in S}{\text{Max}} \left[f(x,y) + \rho \int v_{n-1}(ay+bx-u) dP(u) \right].$$

The analysis of risk for the continuous variable case closely parallels that given in Part Two for discrete variables. If we assume that the range of the random variable is bounded so that the set S is closed and bounded, and that the function f is bounded and continuous, then the various propositions of Part Two may be extended. In particular value and policy iteration may be performed both on the discounted and on the undiscounted problem. In the particular case of the inventory problem equation (2) assumes the form

(3)
$$v_n(y) = \underset{x \geq y}{\text{Min}} \left[k\delta(x-y) + c(x-y) + \right.$$

$$+h\int_0^x (x-u)p(u)\,du + g\int_x^\infty (u-x)p(u)\,du$$

$$+\rho\int_0^x v_{n-1}(x-u)p(u)\,du$$

$$+\rho[1-P(x)]v_{n-1}(0)].$$

Here we have assumed that unsatisfied demand is lost. This value function can be determined by value iteration, but convergence is slow, namely at the rate of interest approximately. It has been shown by SCARF [SCARF (1960)] that with cost functions as assumed in (3) the optimal policy always is (s,S). The value function for the infinite horizon problem may therefore be obtained by solving the integral equation of renewal type obtained by substituting this policy in the right hand side of (3)

$$(4) \qquad\qquad v(y) = k + c(S-y) + v(S) \qquad y \le s$$

$$v(y) = f(y) + \rho[1-P(y)][k+c(S-s)+v(S)] + \rho\int_0^{y-s} v(y-u)p(u)\,du \qquad y \ge s,$$

where we have written

$$(5) \qquad\qquad f(y) = h\int_0^y (y-u)p(u)\,du + g\int_y^\infty (u-y)p(u)\,du.$$

This problem has been treated extensively in the literature [ARROW, HARRIS, MARSCHAK (1951)], [ARROW, KARLIN, SCARF (1958)]. The consumption savings decision and production smoothing and control problems [MILLS (1957)], [BECKMANN (1960)] are of a similar nature.

§ 26. Quadratic Criterion Function

A type of utility and cost function that is very commonly applied is the quadratic. Consider the decision problem under risk in which $f(x,y)$ is a quadratic function of the vector of decision variables x and state variables y, and where x is unrestricted. Assume the principle of optimality in the form (25.2) and let mean and variance of $p(u)$ exist. The principle of optimality is

$$(1) \qquad\qquad v_0(y) = 0,$$

$$v_n(y) = \operatorname*{Max}_x [x'A_n x + 2x'B_n y + y'C_n y + \rho\int v_{n-1}(G_n y + H_n x - u)p(u)\,du$$

where the random variable u may also be a vector (whose dimension equals the row number of G_n and H_n).

Theorem [SIMON (1956)]: *The value function (1) is a quadratic function of the vector of state variables y, and the decision rule is a linear function of the state variables y.*

Proof: For ease of exposition we prove the theorem when x and y are scalar. The generalization to vectors is immediate.

The theorem is trivial for $n=0$. Suppose it is true for $n-1$. By induction hypothesis v_{n-1} is then a quadratic function of $gy+hx-u$

$$= v_0 + v_1(gy+hx-u) + v_2(gy+hx-u)^2 \qquad \text{(say)}.$$

Let

(2)
$$Eu=\mu_1 \qquad Eu^2=\mu_2.$$

Then

$$\int v_{n-1}\, p(u)\, du = v_0 + v_1\, gy + v_1\, hx - v_1\mu_1$$
$$+ v_2 g^2 y^2 + v_2 h^2 x^2 + v_2\mu_2 + 2v_2 ghxy$$
$$- 2v_2 h\mu_1 x - 2v_2 g\mu_1 y.$$

Differentiating the right hand side of (1) with respect to x and setting zero we obtain the optimal decision rule

(3)
$$x = \frac{2\rho v_2 h\mu_1 - \rho v_1 h - 2\rho v_2 ghy - 2by}{2a + 2\rho v_2}$$

which shows x to be linear with respect to the state variable y and to the expected value of the random variable μ.

Substituting for x in the right hand side of (1) and integrating we obtain v_n as a quadratic function of y. QED.

In the single variable case it is instructive to evaluate the stationary v in terms of the coefficients a, b, c, ρ of

(4)
$$v(y) = ax^2 + 2bxy + cy^2 + \rho \int v(x+y-u)\, p(u)\, du$$

$$v_0 + 2v_1 y + v_2 y^2 = \operatorname*{Max}_x \{ ax^2 + 2bxy + cy^2$$

$$+ \rho \int [v_0 + 2v_1(y+x-u) + v_2(y+x-u)^2]\, p(u)\, du \}.$$

Comparing coefficients of y^2, y and 1

$$v_0 = \frac{\rho}{1-\rho}\left[v_2(G^2+\mu^2) - \rho\mu 2v_1 + \frac{(\rho v_1 - \rho\mu v_2)^2}{(1-\rho v_2)} \right],$$

(5)
$$v_1 = \frac{\rho\mu(b+1)v_2}{\rho v_2 + \rho b + \rho - 1},$$

$$v_2 = \frac{1}{2\rho} - b - 1 - \sqrt{\left(1+b-\frac{1}{2\rho}\right)^2 + \frac{1-b^2}{\rho}}.$$

The optimal decision rule is

(6)
$$x = \frac{\rho v_2(y-\mu) + by + \rho v_1}{1-\rho v_2}.$$

Since v_2 is negative the action level x is a decreasing function of the stock level y, and an increasing function of the expected demand μ.

Quadratic criterion functions have been used extensively in the study of production smoothing problems [HOLT, MODIGLIANI, SIMON, MUTH (1960)], and also in macroeconomic planning [THEIL (1964)].

Consider also the allocation of income (or rather wealth) to saving and consumption in the presence of a quadratic utility function

$$(7) \qquad u = u_0 + u_1 c + u_2 c^2$$

and when income is a random variable with mean μ and variance σ^2. If their sign is unrestricted we may conclude that consumption c (or saving) depends linearly on wealth w and on expected income μ

$$c = c_0 + c_1 w + c_2 \mu.$$

Linear consumption functions (with and without risk) have been used widely in economic analysis [SAMUELSON, (1967), 198—209].

Of course, DP can also be used to derive consumption savings function for households and firms also in the non-linear case [cf. PHELPS, et al. (1962)].

Linear prediction. Let us return to the adaptive programming problem of § 18. Under a quadratic utility function, the optimal decision rule was shown to be linear not only in the state variables but also in the expected value of the random variable. Consider its conditional expectation given the past values

$$u_{t-1}, u_{t-2}, \ldots, u_{t-m}, \ldots$$

of the random variable. The problem of the best linear predictor of a random variable from a stationary stochastic process is the classical prediction problem of WIENER [WIENER (1949)]. Suppose for instance that the process is a normal WIENER-BACHELIER process. Let its autocorrelation function be

$$(8) \qquad a_k = \frac{1}{m-k} \sum_{i=1}^{m-k} u_{t-i} u_{t-k-i}$$

and consider the co-variance matrix

$$(9) \qquad A = \begin{Vmatrix} a_0 & a_1 & a_2 & \ldots & a_m \\ a_1 & a_0 & a_1 & & \cdot \\ \cdot & & \ddots & & \cdot \\ \cdot & & & \ddots & \cdot \\ \cdot & & & \ddots & a_1 \\ a_m & & & a_1 & a_0 \end{Vmatrix}$$

and its inverse

$$
(10) \qquad A^{-1} = \left\| \begin{matrix}
b_0 & b_1 & b_2 & \cdots & b_m \\
b_1 & b_0 & b_1 & \cdots & b_{m-1} \\
 & & \cdot & \cdot & \\
\cdot & & & \cdot & \\
\cdot & & & & \cdot \\
\cdot & & & & \\
b_m & \cdot & \cdot & \cdot & b_1 \; b_0
\end{matrix} \right\|
$$

Then the best linear predictor is

$$
(11) \qquad u_t = b_1 u_{t-1} + b_2 u_{t-2} + \cdots + b_m u_{t-m}
$$

where the coefficients b_k are taken from the first row of (10). Note that the estimation can here be performed independently of the choice of action, a convenience achieved through the assumption of a quadratic criterion function with its resulting linear decision rule [cf. MARSCHAK (1963)].

References and Selected Reading to Part Three

ANTOSIEWICZ, H., and A. HOFFMAN: A Remark on the Smoothing Problem. MS 1, 1, 92—95 (1954).

ARIS, R.: The Determination of Optimum Operating Conditions by the Methods of Dynamic Programming. Zeitschrift f. Elektrochemie 65, 3, 229—244 (1961).

ARROW, K. J., T. HARRIS, and J. MARSCHAK: Optimal Inventory Policy. Econometrica 19, 3, 250—272 (1951). Erratum: 20, 1, 133 (1952).

—, S. KARLIN, and H. SCARF: Studies in the Mathematical Theory of Inventory and Production. Stanford: Stanford University Press 1958.

BECKMANN, M.: Production Smoothing and Inventory Control. OR 9, 4, 456—467 (1961).

BELLMAN, R. E.: On the Theory of Dynamic Programming. Proc. Nat. Acad. Sciences, USA 38, 8, 716—719 (1952).

— On the Theory of Dynamic Programming: A Warehousing Problem. MS 2, 3, 272—275 (1956).

— Dynamic Programming. Princeton: Princeton University Press 1957.

—, and St. E. DREYFUS: Applied Dynamic Programming. pp. 104—113. Princeton: Princeton University Press 1962.

— An Application of Dynamic Programming to Location-Allocation Problems. SIAM Review 7, 1, 126—128 (1965).

BURT, OSCAR, R.: Optimal Resource Use over Time with an Application to Ground Water. MS 11, 1, 80—93 (1964).

DANNERSTEDT, G.: Production Scheduling for an Arbitrary Number of Periods given the Sales Forecast in the Form of a Probability Distribution. OR 3, 3, 300—318 (1955).

DANTZIG, G. B.: Linear Programming and Extension. Ch. 23. Princeton: Princeton University Press 1963.

FISHER, J. L.: A Class of Stochastic Investment Problems. OR 9, 1, 53–65 (1961).

GALLIHER, H. P.: Production Scheduling. In: „Notes on Operations Research 1959" assembled by the Operations Research Center, MIT. Cambridge, Mass.: Institute of Technology 1959.

HOLT, C. C., F. MODIGLIANI, and H. A. SIMON: A Linear Decision Rule for Production and Employment Scheduling. MS **2**, 1, 1—30 (1955).

— et al.: Planning Production, Inventories, and Work Force. Englewood Cliffs, New Jersey: Prentice Hall 1960.

HOTELLING, H.: The Economics of Exhaustible Resources. JPE **39**, 137—175 (1931).

JOHNSON, S. M.: Sequential Production Planning over Time at Minimum Cost. MS **3**, 4, 435—437 (1957).

KLEIN, M.: On Production Smoothing. MS **7**, 3, 286—293 (1961).

LITTLE, J. D. C.: The Use of Storage Water in a Hydroelectric System. OR **3**, 2, 187—197 (1955).

LOFTSGARD, L. D., and E. O. HEADY: Application of Dynamic Programming Models for Optimum Farm and Home Plans. Journal of Farm Economics **41**, 1, 51—62 (1959).

LÜTTSCHWAGER, J. M.: Dynamic Programming in the Solution of a Multistage Reliability Problem. Journal of Industrial Engineering **15**, 4, 168—175 (1964).

MARSCHAK, J.: On Adaptive Programming Statistical Control of Time Standards. MS **9**, 4, 137f. (1963).

MILLS, E. S.: The Theory of Inventory Decisions. Econometrica **25**, 2, 222—238 (1957).

MORISHIMA, M.: Prices and the Turnpike. II. Proof of a Turnpike Theorem: the „No Joint Production" Case. Review of Economic Studies **28**, 89—97 (1960—1961).

NEMHAUSER, G. L.: Decomposition of Linear Programs by Dynamic Programming. NRLQ **11**, 2/3, 191—196 (1964).

PHELPS, E. S.: The Accumulation of Risky Capital: A Sequential Utility Analysis. Econometrica **30**, 4, 729—743 (1962).

RADNER, R.: Paths of Economic Growth that are Optimal with Regard only to Final States: A Turnpike Theorem. Review of Economic Studies **28**, 98—104 (1960-1961).

— Notes on the Theory of Economic Planning. Athens: Center of Economic Research 1963.

— Dynamic Programming of Economic Growth. Activity Analysis in the Theory of Growth and Planning (Bacarach and Malinvaud, eds.). Macmillan: London 1967.

ROBERTS, S. M.: Stochastic Models for the Dynamic Programming Formulation of the Catalyst Replacement Problem. Proc. of the Symposium on Optimization Techniques in Chemical Engineering, (G. H. Hsieh, ed.). New York: New York University Office of Special Services to Business and Industry 1960.

SAMUELSON, P. A.: Economics. (7th ed.) New York: McGraw Hill 1967.

SCARF, H. E.: The Optimality of (S, s) Policies in the Dynamic Inventory. In: Mathematical Methods in the Social Sciences, K. J. ARROW, S. KARLIN and P. SUPPES, eds. Stanford: Stanford Univ. Press 1960.

SCHNEIDER, E.: Produktion und Lagerhaltung bei einfacher Produktion. Archiv für Mathematische Wirtschafts- und Sozialforschung, Leipzig. **4**, 1, 99—120 (1938).

SHEPHARD, R. W.: Dynamic Programming. Produktionsplanung zum Zweck der Minimierung der Produktions- und Lagerkosten. Ablauf- und Planungsforschung **5**, 1, 1—25 (1964).

SIMON, H. A.: Dynamic Programming under Uncertainty with a Quadratic Criterion Function. Econometrica **24**, 1, 74—81 (1956).

THEIL, H. et al.: Optimal Decision Rules for Government and Industry. Amsterdam: North Holland Publishing Comp. 1964.

WAGNER, H. M.: Statistical Management of Inventory Systems. New York: Wiley 1962.
WIENER, N.: The Extrapolation, Interpolation and Smoothing of Stationary Time Series with Engineering Applications. New York: Wiley 1960.
ZABEL, E.: Efficient Accumulation of Capital for the Firm. Econometrica 31, 1—2, 131—150 (1963).

Decision Processes in Continuous Time

§ 27. Discrete Action

Consider a system whose output rate decreases with time until corrective action (maintenance, replacement) is taken. Let

t denote the time elapsed since the last action,

$a(t)$ the output rate

$$0 \leq a(t) \leq A.$$

Note that the times of action are regeneration points for the process under consideration: Following these, all previous history is irrelevant for decisions. Let k be the cost of corrective action, assumed to be constant (or constant on the average). The discount factor ρ is now replaced by a continuous discount rate α

$$\rho = e^{-\alpha}.$$

Whatever decision rule is chosen, since all returns (costs) are finite and discounted an upper bound to total value must exist. We may conjecture that a maximum value exists, since the optimal policy is apt to have a simple structure.

As before let v denote this maximum value, the value function. The state variable is obviously the time t since the last action

$$v = v(t).$$

This is perhaps the simplest example of a continuous time decision process with discrete NO-GO action. Let us attempt to formulate the principle of optimality by comparing the system at two closely spaced times t and $t + \Delta t$.

If action is taken at time t then t is made zero at a cost k, and an output rate of $a(0) \Delta t$ is obtained during the following interval. Otherwise the output rate is $a(t) \Delta t$. The remaining returns are given by the function $v(\Delta t)$ and $v(t + \Delta t)$, respectively. Choosing the maximum of the two alternative yields the principle of optimality

(1) $v(t) = \text{Max}\left[-k + a(0) \Delta t + e^{-\alpha \Delta t} v(\Delta t), a(t) \Delta t + e^{-\alpha \Delta t} v(t + \Delta t) \right].$

We assume that $v(t)$ is continuously differentiable with respect to time so that

$$v(t + \Delta t) = v(t) + \Delta t\, v'(t) + o(\Delta t).$$

We also use the expansion

$$e^{-\alpha \Delta t} = 1 - \alpha \Delta t + o(\Delta t)$$

where $o(\Delta t)$ denotes terms of higher order than Δt.

Substituting

$$v(t) = \operatorname{Max}\left[-k + a(0)\Delta t + v(0) - v(0)\alpha \Delta t + v'(0)\Delta t + o(\Delta t),\right.$$
$$\left. a(t)\Delta t + v(t) - \alpha \Delta t\, v(t) + v'(t)\Delta t + o(\Delta t)\right].$$

Considering terms free of Δt one has

$$(2) \qquad\qquad v(t) = \operatorname{Max}\left[-k + v(0), v(t)\right]$$

as the rule for taking action derived from the principle of optimality. Suppose now that no action is to be taken at t. Then the second alternative gives the higher return and

$$v(t) = v(t) + a(t)\Delta t - \alpha v(t)\Delta t + v'(t)\Delta t + o(\Delta t).$$

Cancelling $v(t)$, dividing by Δt and going to the limit $\Delta t \to 0$ one obtains

$$(3) \qquad\qquad v'(t) - \alpha v(t) = -a(t)$$

as a simple differential equation determining v during those times when no action is taken. The principle of optimality—in integrated form—consists therefore of (2) with (3) added as a constraint determining v during the periods of no action.

Multiplying (3) by $e^{-\alpha t}$ we see that

$$(v(t)e^{-\alpha t})' = -a(t)e^{-\alpha t}$$

$$(4) \qquad\qquad v(t) = v(0)e^{\alpha t} - e^{\alpha t}\int_0^t a(\tau)e^{-\alpha \tau}d\tau.$$

The condition for an action to occur at some time $t = T$ is

$$v(0) + k = v(T) = v(0)e^{\alpha T} - \int_0^T e^{\alpha(T - \tau)}a(\tau)d\tau.$$

In that case

$$(5) \qquad\qquad v(0) = \frac{ke^{-\alpha T} + \int_0^T e^{-\alpha t}a(t)dt}{1 - e^{-\alpha T}}.$$

Obviously $v(0)$ depends on T and vice versa $= v(0; T)$. Using (4) $v(t)$ may now be written explicitly

$$(6) \qquad v(t;T) = \int_t^T e^{-\alpha \tau} a(\tau) d\tau + \frac{e^{-\alpha(T-t)}}{1 - e^{-\alpha T}} \left[-k + \int_0^T e^{-\alpha \tau} a(\tau) d\tau \right].$$

It is seen that the assumption of continuous differentiability is indeed satisfied when $a(t)$ is continuous. The next step is obviously to look for the maximizing T. (This corresponds to policy iteration.) Setting the derivative equal to zero we have

$$(7) \qquad \frac{a(T)}{\alpha} + k = \frac{-k e^{-\alpha T} + \int_0^T e^{-\alpha t} a(t) dt}{1 - e^{-\alpha T}} = v(0, T)$$

as a necessary condition for a maximizing T.

This condition may be written in the form

$$(8) \qquad k = \int_0^T e^{-\alpha t} [a(t) - a(T)] dt.$$

From (8) it may be seen that to take action can be optimal only when the output rate $a(t)$ is a decreasing function of time from some point on—which is plausible enough. In the fact the second order conditions (not given here) for a maximum show that at the optimal T, $a'(T) < 0$.

Equations (5) and (7) may be interpreted as follows. The numerator in (5) gives the discounted stream of returns minus costs over an interval between two actions.

$$\frac{1}{1 - e^{-\alpha T}}$$

is the present value of a return coming in at intervals of length T. Both combined give the present value for the system under a cyclic policy of actions equally spaced at intervals of length T. Thus, we encounter again the optimal cycles of Part One.

Upon multiplication by α, equation (7) expresses the economic principle that for a maximal return per unit time the average return $v(0; T)$ must equal the marginal return

$$\alpha k + a(T)$$

obtained through postponing action by an infinitesimal stretch of time. Equation (8) in particular shows that action never pays when the output rate is constant or rising.

The analysis may be extended in a straightforward way to the case of a finite number of discrete actions k, if their effect is to restore the system to an effective operating age τ_k at a cost K_k. The previous analysis properly extended shows that either no action is ever taken or that there exists an earliest critical age $\hat{t}$ such that some action $\hat{k}$ is taken which reduces the operating age to $\tau_{\hat{k}} < \hat{t}$. (Action increasing the operating age is disregarded.) From the definition of v it follows that no waiting for later action can be optimal, and the cycle is then repeated with only one type of action being applied.

It should be noted that these considerations apply on the assumption that the correct value function is used. To find the optimal cycle we may either use policy iteration or compare the minima of expressions (5) under the various actions k.

§ 28. Variable Level

The replacement problem. If the present model is applied to replacement, e. g. for automobiles, only new cars (or cars of a fixed age) would be admitted as replacements. Consider now replacement by cars of any age at a cost $k(\tau, t)$ which depends on the age t of the car being replaced and the age τ of the replacement. The output rate $a(t)$ must now be interpreted as the utility of car ownership minus current operating cost. In principle $\tau > t$ (replacement by an older vehicle) is possible but will be disregarded here. If we rule out actual replacement by a different car of the same age (which does not make sense under certainty) then $k(\tau,t)=0$.

The principle of optimality is now

(1) $$v(t) = \max_{0 \leq \tau \leq t} \left[-k(\tau,t) + a(\tau)\Delta t + e^{-\alpha \Delta t} v(\tau + \Delta t) \right].$$

Expanding $\qquad v(\tau + \Delta t)$ and $e^{-\alpha \Delta t}$

(2) $$v(t) = \max_{0 \leq \tau \leq t} \left[-k(\tau,t) + v(\tau) + a(\tau)\Delta t - \alpha v(\tau)\Delta t + v'(\tau)\Delta t + o(\Delta t) \right].$$

For the maximization of the right hand side only terms not involving Δt are relevant

(3) $$v(t) = \max_{0 \leq \tau \leq t} \left[-k(\tau,t) + v(\tau) \right].$$

At any time t the car owner must consider switching to that age which maximizes the present value (in utility terms) of car ownership.

Suppose t is not a time of replacement. Then the maximum is taken on for $\tau = t$. Dropping $v(t)$ on both sides of (2), dividing by Δt and letting

$$\lim_{\Delta t \to 0} \frac{o(\Delta t)}{\Delta t} = 0$$

we obtain once more equation (27.3)

(4) $$v'(t)-\alpha v(t)=-a(t).$$

Suppose that we start out with a new car and that the first time when $\tau \neq t$ in (3) is at $t=T$. We have the terminal condition

$$v(T)=-k(\tau,T)+v(\tau)$$

where τ is determined as the maximizer of

$$\underset{t}{\text{Max}}[-k(t,T)+v(t)].$$

Write $v(t;\tau,T)$ for $v(t)$ when the parameters τ and T are given. Then

(5) $$v(\tau;\tau,T) = \frac{k(\tau,T)e^{-\alpha(T-\tau)}+\int_{\tau}^{T-\tau} e^{-\alpha(t-\tau)}a(t)dt}{1-e^{-\alpha(T-\tau)}},$$

(6) $$v(t;\tau,T)=\int_{t}^{T}e^{-\alpha(x-\tau)}a(x)dx+e^{-\alpha(T-t)}v(T;\tau,T).$$

It is clear that the optimal policy is again cyclic with parameters τ, T obtained by maximizing (6) with respect to τ and T (policy iteration).

Of course, value iteration is also possible, working backwards in time starting at a terminal time t_0 with $v(t_0)=0$ and solving equation (4) until at $t=t_1$ the principle of optimality requires action. Using the resulting $v(t)$ as terminal value solve the differential equation once more and continue in this way. With efficient programs for solving differential equations this procedure may sometimes be superior to direct maximization of (6).

The optimal τ may or may not be $\tau=0$ depending on the utility function $a(t)$. At the time of replacement $a(t)$ must be decreasing (by virtue of second order conditions).

As BELLMAN has pointed out [BELLMAN (1955)] (see also the example in § 13) when the replacement time T has been exceeded through an oversight of the decision maker then the optimal policy may be to postpone action until some other critical age is reached.

Some *caveat* applies to all replacement problems. Since prices are assumed given the problem as formulated applies at the individual level and not at that of the community as a whole. An interesting question is what would happen if all decision makers had equal preferences or if the same objective output function $a(t)$ was used by everybody. The market value of used cars would then have to adjust itself in such a way that everybody is willing to hold cars for the period which is technically optimal and to replace it then by a factory new vehicle. A person buying a car for the first time would have to be indifferent as to the age of the vehicle he aquires given their market values.

§ 29. Risk of Termination

Let us introduce risk first as the chance that the entire decision process is terminated by an event not under the control of the decision maker.

Processes of this type were introduced by BELLMAN under the heading of Gold Mining. Suppose that of a total available gold stock of 1, an amount x has been extracted and that the current extraction rate is $\lambda(1-x)$, where λ is constant. The risk of termination is indicated by the constant failure rate μ. When the mining operation is stopped without failure having occured the machinery valued at K (in units of gold) may be recovered; otherwise it is lost. The decision is, when to stop mining gold. Other applications (to bombing, search, exploration) are apparent.

Let $v(t)$ denote the total expected future return under the optimal policy when an amount x has been extracted. The principle of optimality sets up a choice among the alternatives of termination and continuation

$$v(x) = \text{Max} \left[K, \lambda(1-x)\Delta t + \mu \Delta t \cdot 0 + (1 - \alpha \Delta t)(1 - \mu \Delta t) v(x + \lambda(1-x)\Delta t) \right].$$

Here the first alternative is to stop mining and recover K. The second involves an expected yield $\lambda(1-x)\Delta t$ plus a risk of failure $\mu \Delta t$; the remainder is the expected discounted return Δt later.

If the answer is to continue mining, then

$$v(x) = \lambda(1-x)\Delta t - (\alpha + \mu)\Delta t\, v(x) + \lambda(1-x)\Delta t\, v'(x).$$

Letting $\Delta t \to 0$ we obtain

$$v'(x) - \frac{\alpha + \mu}{\lambda(1-x)} v(x) = -1.$$

Write

$$\beta = \frac{\alpha + \mu}{\lambda}.$$

Then

$$v'(x) - \beta \frac{v(x)}{1-x} = -1.$$

Its solution

$$v(x) = \frac{1-x}{1+\beta}$$

tells us to stop mining when

$$\frac{1-x}{1+\beta} = K \quad \text{or}$$

$$\lambda(1-x) = (\lambda - \alpha - \mu)K$$

i. e., the rate of recovery of gold must equal the net rate of return on capital K. Of course, mining should never have been started unless (at $x=0$)

$$\frac{1}{1+\beta} > K.$$

§ 30. Discontinuous Processes — Repetitive Decisions

In the following we consider decision processes that are based on events which occur either at random or at intervals whose length is independently and identically distributed. These events may be: failure of equipment, a demand for a commodity (say a repair part), the arrival of a customer, the termination of service to a customer etc.

For concreteness we consider the problem of preventive maintenance of equipment whose reliability function is known. Here the reliability function is defined as the probability that the equipment is still functioning considered as a function of the time since the last maintenance operation or installation. In other words after each maintenance operation (whether it consists of repair or inspection) the equipment is considered to be as good as new.

Let $R(t)$=reliability with $\quad R(0)=1, \quad \dot{R}(t) \leq 0$

where a dot denotes the time derivative. Let

$\mu(t)$ = failure rate defined by

$$(1) \qquad -\frac{\dot{R}}{R} = (-\log \dot{R}) = \mu$$

k = cost of routine preventive maintenance

K = cost of repair of failed equipment. (This cost includes the opportunity cost of the stoppage of operation.)

α = continuous interest rate.

We disregard the duration of repairs and inspections since the effect of down time may be expressed in terms of maintenance cost. Because of the repetitive nature of maintenance this problem is best considered with an unlimited horizon. If no preventive maintenance took place failures would occur at average intervals of length

$$\int_0^\infty t(-\dot{R})\,dt = -Rt\,|_0^\infty + \int_0^\infty R\,dt = \int_0^\infty R\,dt = \tau \quad \text{(say)},$$

provided $\lim_{t\to\infty} tR = 0$ and provided the integral exists. (If it does not exist, expected life is infinite and the expected replacement cost is zero.) Expected discounted maintenance cost is now

$$Ke^{-\alpha\tau} + Ke^{-2\alpha\tau} + \cdots = \frac{Ke^{-\alpha\tau}}{1-e^{-\alpha\tau}}$$

and this is clearly bounded. Now, if the optimal policy gives rise to integrable cost functions then the value function must exist which gives the expected cost of maintenance under the optimal policy. In view of our assumptions the only state variable is the time elapsed since the last inspection. Thus, the principle of optimality reads

$$(2) \qquad v(t) = \text{Min} \begin{cases} k + v(0) \\ \mu(t)\,\varDelta t\,[K + v(0)]\,e^{-\alpha\varDelta t} + [1 - \mu(t)\,\varDelta t]\,e^{-\alpha\varDelta t}\,v(t + \varDelta t). \end{cases}$$

Here the first alternative denotes the cost of immediate preventive maintenance resulting in "new" equipment. In the second alternative failure occurs with probability $\mu(t)\,\varDelta t$ causing a cost K and resulting in new equipment. Otherwise with probability $1 - \mu(t)\,\varDelta t$ the equipment ages by $\varDelta t$ and acquires a value $v(t + \varDelta t)$.

Suppose the second alternative is the minimizing one. Using the Taylor expansions of

$$e^{-\alpha\varDelta t} \quad \text{and} \quad v(t + \varDelta t)$$

we have

$$v(t) = \mu(t)\,\varDelta t\,[K + v(0)(1 - \alpha\,\varDelta t)] \\ + v(t)\,[1 - \mu(t)\,\varDelta t - \alpha\,\varDelta t] + \dot{v}(t)\,\varDelta t + o(\varDelta t).$$

After cancellation of $v(t)$ and division by $\varDelta t$ we obtain the differential equation

$$(3) \qquad \dot{v}(t) - [\alpha + \mu(t)]\,v(t) = -\mu(t)\,K - \mu(t)\,v(0) \\ = -\mu(t)\,c \qquad \text{say,}$$

where

$$(4) \qquad\qquad K + v(0) = c.$$

With the integrating factor $e^{-\alpha t - \int_0^t \mu(\tau)d\tau}$ this differential equation assumes the form

$$\frac{d}{dt}\left(v(t)\, e^{-\alpha t - \int_0^t \mu\, d\tau}\right) = -\mu(t)\, c\, e^{\alpha t + \int_0^t \mu\, d\tau}$$

or

$$v(t)\, e^{-\alpha t - \int_0^t \mu\, d\tau} - v(0) = -\int_0^t c\, \mu(s)\, e^{-\alpha s - \int_0^s \mu\, d\tau}\, ds,$$

(5) $$v(t) = v(0)\, e^{\alpha t + \int_0^t \mu\, d\tau} - c\, e^{\alpha t + \int_0^t \mu\, d\tau}\int_0^t e^{-\alpha s}\, \mu(s)\, e^{-\int_0^s \mu\, d\tau}\, ds.$$

By a well-known formula of reliability theory (in fact the integral of (1) above) we have

(6) $$e^{-\int_0^t \mu\, d\tau} = R(t)$$

(7) $$\mu(t)\, e^{-\int_0^t \mu\, d\tau} = \mu(t)\, R(t) = -\dot{R}(t)$$
$$= q(t)$$

the density function of the life time distribution to failure. With this notation

$$v(t) = \frac{v(0)\, e^{\alpha t}}{R(t)} - \frac{c\, e^{\alpha t}}{R(t)}\int_0^t e^{-\alpha s}\, q(s)\, ds.$$

Now, at $t=0$ clearly

$$v(t) < k + v(0).$$

Thereafter v may or may not increase with time. If it ever reaches the critical level

$$k + v(0)$$

the first alternative will be applied, i.e. preventive maintenance. Let this first happen after time T. Then

$$k + v(0) = v(T),$$

$$= \frac{v(0)\, e^{\alpha T}}{R(T)} - e^{\alpha T}\left[\frac{K + v(0)}{R(T)}\right]\int_0^T e^{-\alpha s}\, q(s)\, ds.$$

8*

Solving for $v(0)$ we have

$$v(0)\left[1 - \frac{e^{\alpha T}}{R(T)} + \frac{e^{\alpha T}}{R(T)}\int_0^T e^{-\alpha s} q(s)\,ds\right] = -e^{\alpha T}\frac{K}{R(T)}\int_0^T e^{-\alpha s} q(s)\,ds - k$$

from which

$$v(0) = \frac{k\,R(T)\,e^{-\alpha T} + K\int_0^T e^{-\alpha s} q(s)\,ds}{1 - \int_0^T q(s)\,e^{-\alpha s}\,ds - e^{-\alpha T} R(T)}.$$

Since

$$\int_0^T q(s)e^{-\alpha s}\,ds = -\int_0^T \dot{R}(s)e^{-\alpha s}\,ds = -R(T)e^{-\alpha T} + 1 + \alpha\int_0^T R(s)e^{-\alpha s}\,ds$$

the denominator may be rewritten and

(8)
$$v(0) = \frac{k\,R(T)\,e^{-\alpha T} + K\int_0^T e^{-\alpha t} q(t)\,dt}{\alpha\int_0^T R(t)\,e^{-\alpha t}\,dt}.$$

Here the numerator expresses the expected discounted cost during a maintenance cycle and the denominator the expected discounted length of that cycle.

This is brought out more clearly when no discount is used. Consider

(9)
$$\bar{v} = \lim_{\alpha \to 0} v(0) = \frac{k\,R(T) + K[1 - R(T)]}{\int_0^T R(t)\,dt}$$

since

$$\int_0^T q\,dt = -\int_0^T \dot{R}\,dt = -R(T) + 1.$$

Here the denominator describes the average length of time between maintenance (preventive or repair) and the numerator the expected cost per cycle. While this formula is plausible enough and might have been guessed, it is not clear in advance that the averaging of cycle time in the denominator and of costs in the numerator can be performed independently as in formula (9).

The optimal T is determined by minimizing the expression (8) for $v(0)$ or (9) for $\bar{v}$, respectively. Differentiating (8) with respect to T and setting zero yields

$$\frac{kR(T)e^{-\alpha T}+K\int_0^T q(t)e^{-\alpha t}\,dt}{\int_0^T R(t)e^{-\alpha t}\,dt} = \frac{k\dot{R}(T)e^{-\alpha T}-\alpha kR(T)e^{-\alpha T}+Kq(T)e^{-\alpha T}}{R(T)e^{-\alpha T}}$$

$$= -k\mu(T)-\alpha k+K\mu(T),$$

or

$$(10)\quad kR(T)e^{-\alpha T}+K\int_0^T q(t)e^{-\alpha t}\,dt=[(K-k)\mu(T)-\alpha k]\int_0^T R(t)e^{-\alpha t}\,dt.$$

In the no-discount case the same formula applies with $\alpha=0$. In the general theory of maintenance it is shown how this equation may be solved in practical cases. In particular when $R(t)=e^{-\mu t}$, i. e. when failure is purely random and no "aging" takes place it never pays to do preventive maintenance since the equipment is at all times as good as new. Equation (10) has then the solution $T=\infty$.

Under a "straight line" reliability function

$$R(t)=1-\beta t, \quad 0\leq t \leq \frac{1}{\beta}$$

and no discounting, formula (10) says

$$k(1-\beta T)+K\int_0^T \beta\,dt = (K-k)\frac{\beta}{1-\beta T}\int_0^T [1-\beta t]\,dt,$$

$$k-k\beta T+K\beta T = \frac{(K-k)\beta}{1-\beta T}\left[T-\beta\frac{T^2}{2}\right]$$

a quadratic equation in T whose positive solution is

$$T = \frac{-k+\sqrt{2kK-k^2}}{\beta(K-k)}.$$

Consider also the case of a piece of equipment containing m parts in parallel which fail at random and have a mean life of $1/\mu$ each. The reliability function is

$$(11)\qquad\qquad R(t)=1-(1-e^{-\mu t})^m$$

and the failure rate

$$\mu(t) = \frac{m\mu(1-e^{-\mu t})^{m-1}}{1-(1-e^{-\mu t})^m}.$$

The average cost per cycle is now

$$\bar{v} = \frac{k+(K-k)(1-e^{-\mu T})^m}{\int_0^T [1-(1-e^{-\mu t})^m]\,dt}.$$

With the variable transformation

$$1-e^{-\mu t} = \xi, \quad 1-e^{-\mu T} = x,$$

$$t = -\frac{1}{\mu}\log(1-\xi),$$

$$dt = \frac{1}{\mu}\frac{1}{1-\xi}\,d\xi,$$

$$\bar{v} = \frac{k+(K-k)x^m}{\frac{1}{\mu}\int_0^x \frac{1-\xi^m}{1-\xi}\,d\xi},$$

one obtains

(12)
$$\bar{v} = \frac{\mu k+\mu(K-k)x^m}{x+\frac{x^2}{2}+\cdots+\frac{x^m}{m}}.$$

The minimum problem is not changed by considering $\dfrac{\bar{v}}{\mu(K-k)}$ instead.
Writing $\dfrac{k}{K-k} = a$ the minimand is

$$\frac{a+x^m}{x+\frac{x^2}{2}+\cdots+\frac{x^m}{m}}.$$

Differentiating and setting the derivative zero yields

(13)
$$\frac{mx^{m-1}}{1+x+\cdots+x^{m-1}} = \frac{a+x^m}{x+\frac{x^2}{2}+\cdots+\frac{x^m}{m}}.$$

This is a polynomial equation which may be solved by standard methods. In the special case $m=2$ we have

$$\frac{2x}{1+x} = \frac{a+x^2}{x+\dfrac{x^2}{2}},$$

$$2x^2 - a - ax - x^2 = 0,$$

$$x^2 - ax = a,$$

$$\left(x - \frac{a}{2}\right)^2 = a + \frac{a^2}{4},$$

$$x_{1,2} = \frac{a}{2} \pm \sqrt{\frac{a^2}{4} + a}.$$

The positive root is unique

$$x = \frac{a}{2} + \sqrt{\frac{a^2}{4} + a}.$$

From this

$$t = -\frac{1}{\mu} \log\left(1 - \frac{a}{2} - \sqrt{\frac{a^2}{4} + a}\right).$$

Preventive maintenance can be advantageous only if $0 < x < 1$.

$$\frac{a}{2} + \sqrt{\frac{a^2}{4} + a}, < 1,$$

$$a < 1 - a,$$

$$\frac{k}{K-k} = a < \frac{1}{2},$$

$$2k < K - k,$$

$$k < \frac{K}{3}.$$

How does it affect the replacement problem of § 27 which was considered under certainty, if we introduce risk?

Let $\mu(t)$ be the risk of complete failure and let the trade-in cost be

$$k(\tau, T) = K + c(\tau) - b(T).$$

Then it may be shown that risk of failure takes the place of the output (or utility) function

$$a(t) \rightarrow a(t) - \mu(t)b(t)$$

and that with this substitution the analysis remains valid.

§ 31. Continuous Time Inventory Control

Inventory control may also give rise to a discontinuous stochastic process and an associated dynamic program. Suppose that inventory decisions may be made at any time, and that demands occur at randomly spaced points in time. Let

$$pr \text{ (time between successive demands} \leq t) = e^{-\lambda t}$$

and let the quantity demanded be one unit at a time. Demand is then described by a pure Poisson process.

With the notation of § 17 consider the system at any time t. If this happens to be a time of a decision we let t be the moment after the decision has been made. The cost functions are as in §17. Assume, however, that delivery requires a time interval of arbitrary but fixed length T. It is natural then to let the value function v contain only those costs which are still subject to control, so that at time t no costs are considered which arise before $t + T$, the earliest date for which we can control costs. If the assumption is made that unsatisfied demand is not lost but is backlogged, then the appropriate state variable on which expected costs after $t + T$ depend is current inventory plus outstanding orders (which will come in before $t + T$) to be called gross stock. Let y now denote gross stock.

To develop the principle of optimality consider events during the time interval

$$(t + T, t + T + \Delta t).$$

The expected storage and shortage cost depends on gross stock minus demand in the interval T. This demand is Poisson distributed

$$(1) \qquad\qquad p_r = \frac{(\lambda T)^r e^{-\lambda T}}{r!}$$

where λ is the rate of demand.

Write

$$(2) \qquad\qquad f(y) = h \sum_{r=0}^{y-1} (y-r)p_r + g \sum_{r=y}^{\infty} (r-y)p_r$$

for the expected storage and shortage cost at time $t+T$, given that gross stock at time t is y. Then during the interval considered an expected cost of

$$f(y)\Delta t.$$

is incurred.

With probability $1-\lambda\Delta t$ no demand arises. With probability $\lambda\Delta t$ a demand occurs and a decision has to be taken subsequent to this change of state. Assume that the ordering cost

$$k+cz$$

for ordering an amount z is charged (to the books) at the time the ordering decision is made. Let x be the gross inventory level chosen. Of course, if $x=y-1$ no order is placed. Disposal may be considered, but we restrict ourselves to $x\geq y-1$. The principle of optimality may now be set down as follows

(3) $v(y)= f(y)\Delta t+(1-\lambda\Delta t)(1-\alpha\Delta t)\,v(y)$
$$+\lambda\Delta t(1-\alpha\Delta t)\,\underset{x\geq y-1}{\text{Min}}\,[k\delta(x-y+1)+c(x-y+1)+v(x)].$$

Cancelling $v(y)$ on both sides, disregarding terms of smaller order than Δt and dividing by Δt we have

$$0= f(y)-(\alpha+\lambda)\,v(y)+\lambda\,\underset{x\geq y-1}{\text{Min}}\,[k\delta(x-y+1)+c(x-y+1)+v(x)],$$

(4) $v(y)=a(y)+\rho\,\underset{x\geq y-1}{\text{Min}}\,[k\delta(x-y+1)+c(x-y+1)+v(x)],$

with

$$a(y) = \frac{f(y)}{\alpha+\lambda} \qquad \rho = \frac{\lambda}{\alpha+\lambda}.$$

Consider first those inventory levels y at which no order is placed. Then (4) reduces to the first order difference equation

(5) $v(y)=a(y)+\rho v(y-1).$

Now, suppose that inventory is falling. Either it never pays to reorder or a level is reached for the first time when ordering is optimal. Let this be s, and let the amount ordered be $D=S-s$.

Through successive substitution in (5)

(6) $v(y)=\rho^{y-s}\,v(s) + \sum_{i=1}^{y-s}\rho^{y-s-i}\,v(s+i).$

The principle of optimality states now that

$$v(s+1)=a(s+1)+\rho k+\rho cD+\rho v(S)$$

or in view of (5)

(7) $v(s) = k + cD + v(S)$.

Write (6) for $y = S$ and substitute (7)

$$v(S) = \rho^D [k + cD + v(S)] + \sum_{i=1}^{D} \rho^{D-i} a(s+i),$$

$$v(S) = \frac{\rho^D [k + cD] + \sum_{i=1}^{D} \rho^{D-i} a(s+i)}{1 - \rho^D}.$$

In view of (5)

(8) $v(s) = \dfrac{k + cD + \sum_{i=1}^{D} \rho^{D-i} a(s+i)}{1 - \rho^D}$.

In this way we have reduced this continuous time inventory problem to the format of § 17.

In particular consider

$$\lim_{\rho \to 1} (1-\rho)\, v(s;\rho) = \lim \frac{1-\rho}{1-\rho^D}\left[k + cD + \sum_{i=1}^{D} \rho^{D-i} a(s+i)\right] = \frac{k + cD + \sum_{i=1}^{D} a(s+i)}{D},$$

λ times this expression is

(9) $\dfrac{k + cD + \dfrac{1}{\lambda} \sum_{i=s+1}^{S} f(i)}{D/\lambda} = \bar{v}$

the cost of the system per unit time, for $1/\lambda$ is the average time that gross inventory level stays constant at level i and D/λ is the average length of an inventory cycle. In formula (8) the numerator gives the present value of a cycle and the denominator the present value of one dollar spent every cycle.

This analysis may be extended to general processes with independent increments [BECKMANN (1961)]. The shifts in the inventory level are then no longer one unit but are random variables. If discrete, the inventory model of § 17 is regained, if continuous, the model of § 25. In the case of a purely discontinuous process the probability distributions for the demand during interval T (required in f) is a compounded distribution of the interval and the quantities of demand. Consider for instance the "stuttering Poisson process", where intervals between demands have exponential distribution and the quantities demanded at one event are geometrically distributed

pr (interval between demands $\leq t$) $= e^{-\lambda t}$,

pr (quantity demanded at one event $= r$) $= (1-q)q^r$.

Total demand i for the commodity during an interval of length T has then the distribution

(10) $$p_r = \sum_{u=0}^{\infty} \left(\frac{\lambda T}{n!}\right)^n e^{-\lambda T} \binom{u+i-1}{i} q^n p^i.$$

This is a Poisson weighted sum of negative binomial terms.

When demand is described by an absolutely continuous process of independent increments, then this must be a WIENER-BACHELIER process [FELLER II, (1966), p. 178]. The formulae developed in this section apply then with a normal distribution replacing the Poisson distribution. In the case of an absolutely continuous process the demand distribution over an interval of fixed length is the normal distribution.

§ 32. Continuous Action — Steady State Problems

Consider now the case that action is not taken at discrete points in time but is applied continuously and that the problem is to determine its optimal intensity. Consider first steady state processes under certainty. The following problem in advertising may be taken as a typical example. Suppose that sales s decline at the rate $-\mu s$ in the absence of advertising action. Let the size of the total market be 1 and assume that while advertising does not affect present customers and in particular cannot prevent their dropping the product, it produces a rate of sales proportional to the unsold market $1-s$ and to the rate of advertising expenditure x

(1) $$\dot{s} = -\mu s + \lambda x(1-s).$$

Here λ is a constant expressing advertising effectiveness. Let the profit function be $\pi(s)-x$. The problem is to maximize discounted profits

(2) $$\int_0^{\infty} e^{-\alpha t}[\pi(s)-x]\,dt$$

subject to (1) as a constraint and subject to the condition that advertising expenditure per unit time cannot exceed a certain rate c

(3) $$0 \leq x \leq c.$$

A discount rate α may (as always) include a termination probability.

Obviously the state of the system is indicated by the sales level s. Let $v(s)$ be the value function and consider the application of advertising at the level x during an interval of length Δt. The principle of optimality states that

$$v(s) = \underset{0 \leq x \leq c}{\text{Max}} \left[-x \Delta t + \pi(s) \Delta t + e^{-\alpha \Delta t} v(s - \mu s \Delta t + \lambda(1-s) x \Delta t) \right].$$

Suppose that $v(s)$ is continuously differentiable and may be expanded. Then

$$v(s) = \pi(s) \Delta t + v(s) - \alpha \Delta t \, v(s)$$

$$+ \underset{0 \leq x \leq c}{\text{Max}} \left[-x \Delta t + (-\mu s \Delta t + \lambda(1-s) x \Delta t) \, v'(s) + o(\Delta t) \right].$$

Cancelling $v(s)$ and dropping Δt we have

(4) $$\alpha v(s) = \pi(s) + \underset{0 \leq x \leq c}{\text{Max}} \left[-x + (\lambda(1-s) x - \mu s) \, v'(s) \right].$$

Suppose that the maximizing level x is neither $x = 0$ nor $x = c$. Since the bracket is a linear function of x this requires that the coefficient of x vanishes

(5) $$-1 + \lambda(1-s) \, v'(s) = 0.$$

As a second equation we have (4) with the x terms dropping out due to (5) and the Max operator disappearing

(6) $$\alpha v(s) = \pi(s) - \mu s \, v'(s).$$

Eliminating $v'(s)$ between (5) and (6)

(7) $$v(s) = \frac{1}{\alpha} \left[\pi(s) - \frac{\mu s}{\lambda(1-s)} \right].$$

In view of our objective function (2) the second term in the bracket must represent advertising expenditure

(8) $$x = \frac{\mu s}{\lambda(1-s)}.$$

In fact upon substitution of (8) in (4) the coefficient of $v'(s)$ vanishes and we have

$$\alpha v(s) = \pi(s) - x = \pi(s) - \frac{\mu s}{\lambda(1-s)},$$

in agreement with (7). We have thus found the steady state solution of the problem. The steady state sales level s is determined (using policy iteration) as the maximizer of the right hand side of (7).

If the coefficient of x in (4) is positive

(9) $$-1+\lambda(1-s)\,v'(s)>0$$

then $x=c$. From (1) one has

$$\dot{s}=-\mu s+\lambda c(1-s)=-(\mu+\lambda c)\,s+\lambda c$$

with the solution

(10) $$s=\frac{\lambda}{\mu+\lambda c}-be^{-(\mu+\lambda c)}.$$

As s follows the path (10) either (9) ceases to apply and the optimal steady state level s is reached or s converges to

$$\frac{\lambda}{\mu+c\lambda}$$

the steady state level of s under the application of $x=c$. These considerations apply *mutatis mutandis* to the case

(11) $$-1+\lambda(1-s)\,v'(s)<0$$

in which $x=0$.

The steady state level for $x=0$ is $s=0$. It is reached under an optimal policy only when (11) applies even for $s=0$ so that

(12) $$\lambda v'(0)<1.$$

This would mean that the effectiveness of advertising $\lambda v'$ even when applied to the entire market falls short of its cost 1, and hence that going out of the market is optimal.

§ 33. The Principle of Optimality in Differential Equation Form

Consider now in more general terms decision processes in continuous time with a continuous payoff resulting from the application of a piecewise continuous decision variable $x=x(t)$. Let the system be in state y. In the following interval Δt a return $f(x,y)\Delta t$ is realized through action x operating on state y. At the same time the system changes its state continuously by an amount

(1) $$\Delta y=g(x,y)\Delta t.$$

Let now an optimal policy $x=d(y)$ be applied resulting in a path $y=y(t)$ which may be called an optimal path Γ.

The principle of optimality may now be stated with respect to optimal paths as follows. If Γ is an optimal path from initial state y_0 to a terminal state y_1 then every segment of this path, say from y_2 to y_3 (Fig. 1) must also be an optimal path.

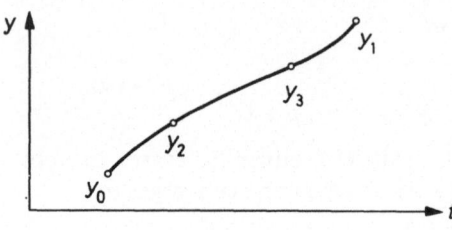

Fig. 1.

If this principle is applied to an infinitesimal segment of the path we have the following version of the principle of optimality

(2)
$$v(y) = \operatorname*{Max}_{\substack{x \\ (x,\,y)\in S}} \left[f(x,y)\Delta t + (1-\alpha\Delta t)v(y+g(y,x)\Delta t) \right].$$

This may be further transformed on the assumption that $v(y)$ is continuously differentiable and that a TAYLOR expansion may be applied

(3)
$$v(y+g(x,y)\Delta t) = v(y) + v'(y)g(x,y)\Delta t + o(\Delta t).$$

Then

$$v(y) = v(y) + \operatorname*{Max}_{\substack{x \\ (x,\,y)\in S}} \left[f(x,y)\Delta t - \alpha\Delta t\, v(y) + v'(y)g(x,y)\Delta t + o(\Delta t) \right]$$

or

(4)
$$\alpha v(y) = \operatorname*{Max}_{\substack{x \\ (x,\,y)\in S}} \left[f(x,y) + v'(y)g(x,y) \right].$$

If v and f are continuously differentiable a necessary condition for maximization of (4) is

(5)
$$f_x(x,y) + v'(y)g_x(x,y) = 0$$

and since this replaces the maximum operator equation (4) assumes the form

(6)
$$\alpha v(y) = f(x,y) + v'(y)g(x,y).$$

The optimal policy $x=x(y)$ and the value function $v(y)$ are then jointly determined by the simultaneous system of differential equations (5), and

(6). In particular if $g(x,y)$ is monotone increasing (or monotone decreasing) in x for every y, then $g(x,y)$ may be chosen as the new variable x, and (5) and (6) assume the simpler form

(7) $$v'(y)= -f_x(x,y),$$

(8) $$xv'(y)-\alpha v(y)=f(x,y).$$

Now, if $f_x(x,y)$ is a montone function of x for every given y the optimum policy is determined by taking the inverse function in (7)

$$x=x(y)= -f_x^{-1}(v'(y),y)=h(v'(y),y), \quad \text{say}$$

and substituting in (8). In this way one obtains the ordinary differential equation in $v(y)$

(9) $$h(c'(y),y)v'(y)-\alpha v(y)= -f(h(v'(y),y))=0.$$

As an example consider the problem of maximizing discounted future profits (i.e. the value of the firm) when demand depends both on price y and of the rate of change of price $x=\dot{y}$. Let q be the quantity demanded and assume

$$q=q(x,y)=a + \frac{bx^2}{y} - cy.$$

This means that demand is a decreasing function of price and an increasing function of the rate of change of price. Since price increases stimulate sales through the expectation of further price increases we want

$$b>0, \quad c>0.$$

Moreover, assume all costs to be fixed, i. e. constant. The profit function is now

$$f(x,y)=yq(x,y)=ay+bx^2-cy^2.$$

Observe that

$$-f_x=-2bx =v'(y)$$

by (7).

Solving for x we have

(10) $$x(y)= -\frac{v'(y)}{2b}.$$

Substituting in (8)

$$\frac{-v'(y)}{2b}v'(y)-\alpha v(y)=ay+b\left(\frac{v'(y)}{2b}\right)^2-cy^2$$

or

(11) $$\frac{3v'^2(y)}{4b} + \alpha v(y) = -ay + cy^2.$$

This (Riccati) differential equation has the solution

$$v = v_0 + v_1 y + v_2 y^2$$

whose coefficients are determined by substitution and termwise comparison

$$\frac{3(v_1 + 2v_2 y)^2}{4b} + \alpha(v_0 + v_1 y + v_2 y^2) = -ay + cy^2,$$

$$\frac{3v_1^2}{4b} + \alpha v_0 = 0,$$

$$\frac{3v_1 v_2}{b} + \alpha v_1 = -a,$$

$$\frac{3v_2^2}{b} + \alpha v_2 = c.$$

Substituting in (7)

$$2v_2 y + v_1 = -f_x(x, y) = -2bx$$

from which we obtain a linear decision rule (as in the case of period problems with a quadratic criterion function)

(12) $$x = -\frac{v_1}{2b} - \frac{v_2}{b} y.$$

We turn now to the general problem and consider the case of a finite horizon. Then the remaining time of the process enters as a further state variable t (notice that t like n before is decreasing in real time). The principle of optimality now states that

(13) $$v(y,t) = \text{Max}\left[f(x,y)\Delta t + (1 - \alpha \Delta t)v(y + g(x,y)\Delta t, t - \Delta t)\right]$$

with a terminal payoff $b(y)$ we have the terminal condition

(14) $$v(y,0) = b(y).$$

If v is assumed continuously differentiable then a TAYLOR expansion leads to

(15) $$\alpha v(y,t) + v_t(y,t) = \underset{\substack{x \\ (x,y) \in S}}{\text{Max}}\left[f(x,y) + v_y(y,t)g(x,y)\right]$$

in particular if $\alpha=0$ we have

$$v_t(y,t)=\underset{(x,y)\in S}{\underset{x}{\text{Max}}}\left[f(x,y)+v_y(y,t)g(x,y)\right].$$

Maximization of the right hand side in (15) requires that

(16) $$f_x(x,y)+v_y(y,t)g_x(x,y)=0.$$

The maximum operator may now be dropped in (15) yielding

(17) $$\alpha v(y,t)+v_t(y,t)=f(x,y)+g(x,y)v_y(y,t).$$

   Equations (16) and (17) are the statement of the principle of optimality which was first obtained by BELLMAN as an alternative formalism in the calculus of variations. (In BELLMAN's case $\alpha=0$). It may be shown that the characteristics of the system of partial differential equations (5), (6) are solutions to the EULER equation of the classical calculus of variations. However, a more direct proof is available that the principle of optimality of DP will also generate the EULER equation (see § 34).
   Again if the function $g(x,y)$ is monotone in x for every y then g may be chosen as the activity variable and the principle of optimality (15) simplifies to

(18) $$\alpha v(y,t)+v_t(y,t)=\underset{(x,y)\in S}{\underset{x}{\text{Max}}}\left[f(x,y,t)+xv_y(y,t)\right]$$

with similar simplifications in (16) and (17). If in addition $\alpha=0$

(19) $$v_y(y,t)=-f_x(x,y,t),$$

(20) $$v_t(y,t)-xv_y(y,t)=f(x,y,t).$$

§ 34. Dynamic Programming and the Calculus of Variations

   Consider the standard problem in the calculus of variations of determining

(1) $$\text{Min}\int_{x_0}^{a}f(x,y,y')\,dx$$

with arbitrary initial conditions $y(x_0)=y_0,$
and given terminal conditions $y(a)=b.$

Suppose that the minimum (1) exists over a suitable class of functions, say the class of continuous functions y with piecewise continuous first and second derivatives y' and y''. Denote the minimum (1) by $v(x_0, y_0)$ or for simplicity $v(x, y)$.

The principle of optimality may now be applied to an initial small displacement $\Delta x, \Delta y$

(2) $$v(x, y) = \underset{y'}{\text{Min}} \left[f(x, y, y') \Delta x + v(x + \Delta x, y + \Delta y) \right].$$

Expanding

$$v(x + \Delta x, y + \Delta y) = \underset{y'}{\text{Min}} \left[f(x, y, y') \Delta x + v(x, y) + v_x(x, y) \Delta x + v_y(x, y) y' \Delta x) \right]$$

or

(3) $$0 = \underset{y'}{\text{Min}} \left[f + v_x + v_y y' \right]$$

where v_x and v_y are independent of this initial y'.

The first order condition for a minimum of the right hand side is

(4) $$v_y = -f_{y'}.$$

Hence from (3)

(5) $$v_x = -f + y' f_{y'}.$$

Equations (4) and (5) give the gradient of the potential or value function $v(x, y)$. They are obtained from the general differential formula

$$dv = f(x, y, p(x)) \left[dx + dy - p(x, y) dx \right] f_p(x, y, p(x))$$

by setting $p = y'$. Notice that this determines an "extremal"

$$p(x, y) = y'.$$

Observe now that

$$f = f(x, y, y'(x, y))$$

and

$$\frac{\partial}{\partial x} v_y = \frac{\partial}{\partial y} v_x.$$

We have

(6) $$\frac{\partial}{\partial x} v_y = \frac{d}{dx} \left[-f_{y'}(x, y, y'(x, y)) \right] \Big|_{y = \text{const.}}$$

$$= -f_{xy'} - f_{y'y'} y'' \Big|_{y = \text{const.}}$$

Also

(7) $\qquad \dfrac{\partial}{\partial y} v_x = \dfrac{d}{dy} \left[-f(x,y,y'(x,y)) + y'(x,y)(f_{y'}(x,y,y'(x,y))) \right] \Big|_{x=\text{const.}}$

$\qquad\qquad = -f_y - f_{y'} \dfrac{dy'}{dy} + \dfrac{dy'}{dy} f_{y'} + y' f_{y'y} + y' f_{y'y'} \dfrac{dy'}{dy} \Big|_{x=\text{const.}}$

Equating (6) and (7) and changing signs we have

$$f_y = f_{xy'} + f_{yy'} \dfrac{dy}{dx} + f_{y'y'} \left[\dfrac{\partial y'}{\partial x} + \dfrac{\partial y'}{\partial y} \dfrac{dy}{dx} \right]$$

and so

(8) $\qquad\qquad\qquad\qquad f_y = \dfrac{d}{dx} f_{y'} .$

This is the EULER equation of the calculus of variations.

If only terminal conditions are fixed equation (4) implies the "free" boundary condition

(9) $\qquad\qquad\qquad\qquad f_{y'} = 0.$

If initial conditions are fixed and terminal conditions are free the process of variation may be performed at the terminal point in the same way yielding (9) as the free terminal condition.

DP may thus be used to solve variational problems through the application of the principle of optimality. This means that value iteration (2) can be performed on $v(x,y)$ directly starting with the terminal condition

$$v(a,b) = 0.$$

This procedure is both direct and relatively simple. It has been used to calculate minimum time-to-climb problems [CARTAINO-DREYFUS (1957)]. Of great advantages is the fact that inequality constraints such as

(10) $\qquad\qquad\qquad c_1(x,y) \leq y' = c_2(x,y)$

are easily introduced in the direct maximization of v.

Policy iteration is also possible in principle. Assume a decision rule

(11) $\qquad\qquad\qquad\qquad y' = d(x,y)$

solve this differential equation and calculate $v(x,y)$. The minimizer $y'(x,y)$ of the right hand side of (2) then produces a new decision rule and the process is repeated. The difficulties of this method lie in the necessity of solving differential equations of type (11).

In fact, several approaches are possible to most economic decision problems. If the solutions are not identical or mathematically equivalent,

9*

at any rate they should be economically similar. Identical solutions are obtained for instance in the case of the linear programming and dynamic programming approaches to determining shortest—or longest—paths in a network, and in fact for the analysis of all deterministic problems involving a finite number of states. Another case in point is the inventory problem where it does not matter whether one determines first the steady state probabilities and optimizes next, or whether the somewhat more general policy iteration method of dynamic programming is applied. Returning to decision problems in continuous time, we observe that the calculus of variations—whenever applicable—the maximum principle of PONTRYAGIN, and the dynamic programming method all yield solutions which are mathematically identical. Dynamic programming seems to be particularly suited however for numerical calculations by direct maximization.

§ 35. Variation under Constraints: The Maximum Principle

We consider the simplest case of a constrained maximum problem in the calculus of variations as formulated in control theory.

$$(1) \qquad \mathop{\mathrm{Max}}_{x,u} \int_t^T f(x,u)\,dt$$

subject to

$$(2) \qquad \dot{x} = g(x,u)$$

the terminal condition,

$$(3) \qquad x(T) = x_1$$

and subject to certain restrictions on the control variable u

$$(4) \qquad u \in U$$

which we leave unspecified.

Since it is in the general spirit of the Dynamic Programming approach we have assumed (as before in § 34) that terminal conditions are prescribed rather than the usual initial conditions. The translation to given initial conditions is immediate.

The value of the maximum is now a function of the arbitrary initial values of x and t. They are the state variables of the Dynamic Program. Write

$$(5) \qquad \mathop{\mathrm{Max}}_{x,u} \int_t^T f(x,u)\,dt = v(x(t),t)$$

subject to (2), (3), (4).

The principle of optimality states that:

(6) $$v(x(t),t) = \underset{x(t+\varDelta t),u}{\text{Max}} \left[f(x,u)\varDelta t + v(x(t+\varDelta t),t+\varDelta t) \right]$$

subject to (2), (4).

Using a Taylor expansion for v we have

(7) $$v(x(t),t) = \underset{\dot{x},u}{\text{Max}} \left[f(x,u)\varDelta t + v(x(t),t) + v_1 \dot{x}\varDelta t + v_2 \varDelta t \right]$$

subject to (2), (4), where we have written

$$v_1 = \frac{\partial v}{\partial x}, \quad v_2 = \frac{\partial v}{\partial x}.$$

Cancelling and reordering terms in (7) we obtain

$$-v_2 = \underset{u,\dot{x}}{\text{Max}} \left[f(x,u) + v_1 \dot{x} \right]$$

or

(8) $$-v_2 = \underset{u \in U}{\text{Max}} \left[f(x,u) + v_1 g(x,u) \right] \quad \text{using (2).}$$

$$= f(x,u^*) + v_1 g(x,u^*), \quad \text{say.}$$

Write $v_1 = \lambda$.

In view of

$$\frac{\partial v_1}{\partial t} = \frac{\partial v_2}{\partial x}$$

we have

(9) $$\frac{\partial \lambda}{\lambda t} = -\frac{\partial}{\partial x} \left[f(x,u^*) + \lambda g(x,u^*) \right].$$

When u is subject to no further constraint, a necessary condition for the maximum in (8) is that

(10) $$\frac{\partial}{\partial u} \left[f(x,u) + \lambda g(x,u) \right] = 0.$$

Equations (9) and (10) are the so-called EULER-LAGRANGE conditions for the constrained maximum problem (the LAGRANGE problem) in the calculus of variations, i. e., the EULER equations as applied to the modified maximand

$$\int_t^T f(x,u) + \lambda(g(x,u) - x)dt$$

where λ is a LAGRANGE multiplier function.

If the control variable is subject to constraints we have instead the maximum principle of control theory. Define a HAMILTONIAN

$$H = \Psi_0 f(x,u) + \Psi_1 g(x,u)$$

in terms of which

(2)
$$\dot{x} = \frac{\partial H}{\partial \Psi_1} = g(x,u),$$

(11)
$$\Psi_0 \equiv 1$$

(12)
$$\dot{\Psi}_1 = -\frac{\partial H}{\partial x} = -\frac{\partial f}{\partial x} - \Psi_1 \frac{\partial g}{\partial x}$$

and let

(13)
$$H = \operatorname*{Max}_{u \in U} H(x, \Psi, u).$$

With $\Psi_1 = \lambda = v_1$ the conditions (11), (12), (13) are identical with (8) and (9) which were derived above. The system (11), (12), (13) is the maximum principle of control theory [PONTRYAGIN, p. 19; cf. also NEMHAUSER, p. 237] for the problem (1), (2), (3), (4) of this chapter.

If the initial value of x may be freely chosen we have furthermore

(14)
$$0 = \frac{\partial v}{\partial x} = v_1 = \lambda(t_0) = \Psi(t_0)$$

as the "free initial condition". While a full discussion of control theory lies outside the scope of this treatise, it is instructive to see how the basic control problem fits into the general framework of dynamic programming. Specifically, the principle of optimality allows an easy and straightforward derivation of the maximum principle. The argument developed here applies in reverse. Thus, the maximum principle of control theory and the principle of optimality of dynamic programming are equivalent. The EULER-LAGRANGE conditions of the calculus of variations are a special case of either.

Mathematical equivalence is not tantamount to equal usefulness in practical application. A great convenience of the dynamic programming approach is the ease of formulating the principle of optimality and the fact that the maximization required by the principle of optimality may be performed directly. Disadvantages arise when the number of state variables is large. On the other hand, the classical calculus of variations sometimes generates explicit solutions, and control theory permits an analysis in terms of phase diagrams.

While the tool kit of the operations researcher contains many tools, it is the art of the analyst to apply the optimal ones—no hard and fast rules can be given here.

References and Selected Reading to Part Four

BARLOW, R. E., F. PROSCHAN, and L. C. HUNTER: Mathematical Theory of Reliability. London: Wiley 1965.

—, and L. HUNTER: Optimum Preventive Maintenance Policies. OR **8**, 1, 90—100 (1960).

BECKENBACH, E. F. (ed.): Modern Mathematics for the Engineer. 2nd series. New York: McGraw Hill 1961.

BECKMANN, M., and R. MUTH: An Inventory Policy for a Case of Lagged Delivery. MS **2**, 2, 145—155 (1956).

— An Inventory Model for Arbitrary Interval and Quantity Distributions of Demand. MS **8**, 1, 35—57 (1961).

BELLMAN, R. E.: Equipment Replacement Policy. SIAM **3**, 133—136 (1955).

— Dynamic Programming. Ch. XII, 245—282. Princeton: Princeton University Press 1957.

— Some Applications of the Theory of Dynamic Programming. A Review. OR **2**, 3, 275—288 (1954).

—, I. L. GLICKSBERG, and O. A. GROSS: Some Aspects of the Mathematical Theory of Control Processes. Santa Monica: RAND Corp. R-313, 1958.

—, and St. E. DREYFUS: Applied Dynamic Programming. Ch. V, 180—206. Princeton: Princeton University Press 1962.

— (ed.): Mathematical Optimization Techniques. Berkeley and Los Angeles: University of California Press 1963.

—, and R. KALABA: Dynamic Programming and Modern Control Theory. New York: Academic Press 1965.

CARTAINO, T. F., and S. E. DREYFUS: Application of Dynamic Programming to the Airplane Minimum Time-to-climb Problem. Aeronautical Engineering Review **16**, 6, 74—77 (1957).

DESCAMPS, R.: Remplacement et Programmation Dynamique. Métra **4**, 2, 181—199 (1965).

DREYFUS, S. E.: A Generalized Equipment Replacement Study. SIAM **8**, 3, 425—435 (1960).

— Dynamic Programming and the Calculus of Variations. New York–London: Academic Press 1965.

FELLER, W.: An Introduction to Probability Theory and its Applications. II, p. 178. New York: Wiley 1966.

JORGENSON, D. W., McCALL and RADNER: Optimal Replacement Polity. Amsterdam: North Holland Publ. 1967.

MAFFEI, R. B.: Planning Advertising Expenditures by Dynamic Programming Methods. Management Technology **1**, 2, 94—100 (1960).

MEIJ, J. L.: Depreciation and Replacement Policy. Amsterdam: North Holland Publishing Comp. 1961.

PONTRYAGIN, L. S. et al.: The Mathematical Theory of Optimal Processes. New York: Wiley 1962.

RADNER, R., and D. W. JORGENSON: Opportunistic Replacement of a Single Part in the Presence of Several Monitored Parts. RAND Corp., RM – 3057 – PR, Nov. 1962.

WEISS, G. H.: On the Theory of Replacement of Machinery with a Random Failure Time, NRLQ **3**, 279—293 (1956).

WHITE, D. J.: Optimal Revision Periods, Journal Math. Anal. Appl. 4, 353—365 (1962).

Author Index

Antosiewicz, H. 104
Aris, R. IX, 104
Arrow, K. J. 61, 64, 83, 101, 104

Barlow, R. E. 135
Bass, H. G. X
Beckenbahe, E. F. 135
Beckmann, M. J. 11, 34, 35, 54, 83, 101,
    104, 122, 135
Bellman, R. E. IX, X, 11, 14, 35, 83, 85,
    104, 111, 112, 129, 135
Blackwell, D. 6, 36
Breiman, L. 83
Burt, O. R. 104

Cartaino, T. F. 131, 135
Cruon, R. IX

Dantzig, G. B. IX, 8, 36, 94, 100, 104
De Cani, J. S. 83
Descamps, R. 135
Donnerstedt, G. 104
Dreyfus, S. E. IX, X, 36, 83, 104, 135
Dubins, L. E. 83

Elsner, K. IX
Everett, H. 6, 36

Feller, W. 72, 83, 123, 135
Fisher, J. L. 104
Fleming, W. H. 83
Friedman, L. X

Galliher, H. P. 105
Ghellinck, G. de 83
Gibson, J. E. IX
Gilford, D. M. 84
Glicksberg, I. L. 135
Gross, O. A. 35, 135

Hadley, G. IX, 75, 83
Harris, T. 61, 83, 101, 104
Heady, E. O. 105
Held, M. 36
Hochstädter, D. 83

Hoffman, A. 104
Holt, G. G. 105
Hotelling, H. 85, 105
Howard, R. A. X, 48, 55, 83
Hunter, L. C. 135

Iglehart, D. L. 83

Jewell, W. S. 83
Johnson, S. M. 95, 105
Jorgenson, D. W. 135

Kalaba, R. 135
Karlin, S. 64, 83, 101, 104
Karp, R. M. 36
Kaufmann, A. IX
Kimball, G. E. X
Kirby, D. J. X
Klein, M. 105

Laderman, J. 11, 35
Lindley, D. V. 83
Little, J. D. C. 96, 105
Loftsgard, L. D. 105
Lüftschwager, J. M. 105

Maffei, R. B. 135
Maitra, A. 83
Manne, A. S. 83
Marschak, J. 61, 83, 101, 104, 105
McCall, J. J. 135
Meij, J. L. 135
Mills, E. S. 101, 105
Modigliani, F. 105
Morgenstern, O. 5, 36, 84
Morishima, M. 105
Murphy, R. E. 83
Muth, R. 135

Nemhauser, G. L. X, 3, 15, 36, 100, 105, 134
Neumann, J. von 5, 36, 84
Noton, A. R. M. X

Phelps, E. S. 105
Piehler, J. X

Pollack, M. 36
Polya, G. 15, 36
Pontryagin, L. S. 132, 134, 135
Proschan, F. 135

Radner, R. 11, 36, 91, 105, 135
Roberts, S. M. 75, 84, 105

Sadowski, W. 36
Samuelson, P. A. X, 101, 105
Sasieni, M. X, 84
Savage, L. J. 83
Savage, R. 36
Scarf, H. E. 64, 79, 83, 84, 101, 104, 105
Schneider, E. 105
Shelley, M. W. 84
Shephard, R. W. 105
Simon, H. A. 101, 105

Simpson, M. G. X
Smith, C. S. X

Theil, H. 105

Wagner, H. M. 75, 84, 105
Wald, A. 1, 36
Weiss, G. H. 135
Wentzel, J. S. X
White, D. J. 135
Whitin, T. M. 75, 83, 84
Wiebenson, W. 36
Wiener, N. 103, 106
Wolfe, P. 100

Yaspan, A. X

Zabel, E. 106
Zschocke, D. 84

Subject Index

action, continuous 123–125
—, discrete 107
—, finite number of discrete 110
—, optimal 109
activity variable 129
adaptive programming 76–81
— —, problem 103
additive constant 25, 45, 48
advertising 123
—, effectivness 123, 125
—, expenditure 124
allocation 85–87
— of income 103
—, problem 93, 87
approximation 65, 75
—, achieved in value iteration, degree 53
—, errors 52
—, Stirling's approximation 81
arrival of a customer 43
automobile replacement 26–33, 110
average payoff 49, 59
—, direct calculation 60
—, maximum 24, 25
— — per decision 24
—, stationary 46
—, total 34
average return 24
— — per transition 33
— —, stationary 43

backward induction 17
Bayesian approach 5
—, policy, optimal 79
Bayes' principle 76, 81
Beta distribution 76, 77
binomial distribution 75, 81
— —, negative 72, 77
— family 77
— —, negative 75, 76
— terms, Poisson weighted sum of 123
bombing 112
boundary condition 79
— —, "free" 131
boundedness 91
Brouwer fixed point theorem 97

calculus of variation 129–132
capital, rate of return 113
chess 14
combinations 12
combinatorial 9
computing capabilities 82
concavity 92, 99
— of the production function 97
cone 92
constraints 99
—, inequality 131
consumption 85, 86
—, initial 93
—, saving decisions 101
—, stationary 93
continuous action 123–125
— differentiability 109
— interest rate 43
— variable case 100
continuous decision variable 85–106
control theory 132
—, maximum principle of 133, 134
convergence 21
—, uniform 90
convexity 92
cost average 2, 72
— of being in the next stage 26
—, carrying 80
— of the decision 26
— of inventory control, total expected
    63, 114
—, minimum 95
—, present value 70
— of repair 113
— of routine preventive maintenance
    113
— of shortage 62, 72, 121
—, shortage penalty 80
— of storage 62, 72, 121
—, reordering and stock-outs 61
cost function, integrable 114
— —, storage and shortage 65
critical age 111
critical level 115
cycle 7, 23, 33

cycle, length of the 35
—, maintenance 116
—, optimal 25
—, present value of a 122
—, time, averaging of 116
cycling 9

dam operation 96
decision, average payoff per 24
—, chain 5, 15, 51
—, classification 5
—, consumption savings 101
—, cost 26
—, depending on the horizon 22
—, in time; sequential 6
—, optimal 46
—, program 15
—, repetitive 113–120
—, sequential 76
—, tree 3, 4, 7
—, unique optimal initial 21
—, variable 87, 99
decision problem, discounted infinite
    horizon 18
— —, infinite horizon 19, 90
— —, sequential 3, 61, 87
— — with finite alternatives in infinite
    horizon, sequential 24
— — with finitely many alternatives,
    undiscounted 25
decision process 1, 55
— — in continuous time 107–134
decision rule 11, 19, 48, 49, 131
— —, change in the 58
— —, improvements in the 58
— —, linear 101, 104, 128
— —, optimal 18, 25, 49, 102
decomposition principle 100
demand 62, 127
—, backlogged 120
— distribution, geometric 72–76, 79, 82
— for a commodity 113
—, non stationary 95
—, unsatisfied 101
differential equation
— —, ordinary 127
— — form, principle of optimality 125
— —, Ricatti 128
— —, simultaneous system of 126
discount 6, 15, 116
— factor 21, 53, 64, 85, 107
discounted future profits 127

discounted maintenance cost, expected
    114
— profits 123
— stream of returns minus cost 109
discrete alternatives 35
distribution, a priori probability 76, 79
—, a posteriori 77
—, Beta 76, 77
—, binomial 75, 81
—, demand 72–76, 79, 82
—, geometric 78
—, negative binomial 72
—, Poisson 76
— to failure, life time 115
—, uniform 77
—, uniform a priori 78
dual variable 34, 55
dynamic program under risk and uncer-
    tainty 61
dynamic programming 33–35
— and the calculus of variations
    129–132

economic growth model 85
eigenvalue 40
eigenvector 40
efficiency price 34, 55
electricity generation 96
equation, difference 39
—, differential 108, 111, 114
—, Euler 129, 131
—, functional 1
—, linear 19
—, inventory 67
— of renewal type, integral 101
— system, linear 48
—, value determination 80
ergodic process 71
ergodicity 40
error, approximation 52
— maximum 53
Euler equation 129, 131
Euler-Lagrange condition 133, 134
expected discounted earnings 41
— discounted maintenance cost 114
— life 114
— replacement 114
exploration 112
exponential weighting 81–82

failure of equipment 113
—, purely random 117

failure rate 112, 113
feasibility 95, 96
feasible 10
— solution 34
finite alternatives 1–35
flow constraint 34
function, autocorrelation 103
—, bi-linear 54
—, concave 97, 98
—, consumption 103
—, expectation 81
—, integrable cost 114
—, Kronecker 63
—, limit 98
—, linear constraint 94
—, loss 63, 69, 74
—, payoff 87
—, piecewise linear 95, 96, 99
—, production 96, 97
—, quadratic criterion 82, 101–104
—, storage and shortage cost 65
—, "straight lim" reliability 117
— technique, generating 67, 72
—, utility 86, 96, 103, 120

game 14
—, state of the 14
—, value of the 14
— with perfect information 14
geometric interpretation 6–10
gold mining 112
gradient of the potential 130
graph 7
growth rate 99

homogeneity 92
—, positive 92
horizon 2, 15, 78, 87, 94
—, decision depending on the 22
—, fixed 88
—, infinite 17, 18, 19, 82, 99
—, planning 65
— problem, infinite 21, 24, 27, 52, 54, 90
—, unlimited 44, 114

improvements, impossibility of further 49
indivisible 9, 11
initial conditions 129, 132
— —, "free" 34
interest rate, continuous 43

inventory equation 67
—, starting 66
inventory control 61–76
— —, continuous time 120–123
— —, expected discounted cost 63
— — without fixed ordering cost 79–81
inventory level 121
— —, shifts 122
inventory problem 82, 100, 132
— —, economics of the 76
irrigation 96

Lagrange function 34
— multiplier 85, 133
limit, finite 43
— function 98
— process 52
linear inhomogeneous problems 94
linear prediction 103
— —, best 103
linear program 33, 34, 95, 99
— —, decomposing a complex 99
linear programming 8, 9, 33–35, 54–55, 132
loss function 63, 64, 74

machine care 55–61
macroeconomic planning 103
maintenance 107
— cost, expected discounted 114
— cycle 116
—, preventive 113, 119
Markov chain 37, 44
— —, cyclic 55
— —, ergodic 45
— process 100
matrix, co-variance 103
—, permutation 19
—, positive 40
—, root of the 40
—, stochastic 38
maximizing actions 18
maximum error 53
maximum problems in recursive form 1
maximum principle 132–134
— — of control theory 133, 134
— — of Pontryagin 132
mean 101
memory 15
mixed strategies 14
modes of transportation 13

monotonicity 91
move 14

neighborhood sets 10
network 7, 8, 34
—, finite 24
von Neumann path 99
newsboy formula 82
— rule 81
no discounting 70–72
non-ergodic problem 55

operating cost 26
— age 110
optimal policy 51
— —, nature 23
— —, structure 66
ordering cost 62, 80, 82
— —, fixed 66
— —, inventory control without fixed 79
— —, unavoidable variable 64

past experiences 37
path in a network, shortest 8
—, von Neumann 99
—, optimal 125, 126
—, shortest 7, 13, 14, 35
payoff 1, 45, 56
— of an action 37
—, average 24, 25, 34, 46, 49, 59, 60
—, continuous 125
—, expected 54, 100
—, one period 79
—, function 87
—, terminal 10, 15, 17, 46, 96
periodicity 96
periods, number 78
permutation matrix 19
plane assignment 11, 12
player 14
Poisson distribution 76
— process 120
— —, stuttering 122
— weighted sum of negative binomial
   terms 123
policy, cyclic 109, 111
—, initial 27
— iteration 19–21, 25, 48–51, 59–60,
   100, 109, 131
—, optimal inventory 63
—, — storage 94
—, s S 65–70, 101

policy, stationary 52, 86
— of stocking a commodity 61
principle of optimality 5, 10–15, 17, 18,
   19, 27, 45–48, 61, 63–65, 78–79, 81, 85,
   88, 90, 91, 95, 96, 99, 100, 101, 107,
   108, 110, 114, 120, 121, 124, 126, 128,
   129, 130, 133
— — in differential equation form
   125–129
probability of a demand for i items in n
   periods 71
— distribution 64
— — a priori 79
— —, priory subjective 76
—, joint 54
— of i events 68
— of shortages 62
—, state 39, 58
—, stationary state 40, 49
—, steady state 72
—, termination 123
—, transiton 37, 38, 45, 56, 58,
   76
problem, adaptive programming 102
—, allocation 93
—, infinite horizon 21, 24, 27, 90
—, inventory 76, 82, 100, 132
—, knapsack 2, 35
—, linear inhomogeneous 94–95
—, maintenance 35
—, minimum time to climb 131
—, non ergodic 55
—, non stationary 91
—, n period 88
—, production smoothing 103
—, — — and control 101
—, replacement 110
—, sequential decision 87
—, shortest path 35
—, stationary 88
—, steady state 123
—, undiscounted 24, 51
—, with growth, undiscounted allocation
   87
—, with horizon n 94
—, without discount and with infinite horizon 24, 52, 55
process, controlled 27
—, discontinuous 113–120
—, ergodic 71
—, limit 52

process, stationary stochastic 103
—, stochastic decision 55
—, "stuttering Poisson" 122
production function 96, 97
— —, homogeneous of degree one 96
—, overtime 95
—, smoothing and control problems 95,
  101, 103

quadratic criterion function 82,
  101–104

random variable 78
rate, continuous interest 113
—, failure 112–113
— of improvement 53
— of sales 123
recursion 11
recursive, definition 41
recursiveness, idea of 61
regeneration points 107
reliability 113
replacement 107
—, expected 114
reservoir 96
resource 99
return 41
—, discounted 55, 109, 112
—, discounted and finite 107
—, expected 41, 46, 112
—, — one period 57
— on capital, rate of 113
—, stationary average 43
—, undiscounted future 42
risk 5, 37–84, 100–101, 119
—, dynamic program under 61
— of termination 112–113

search 112
sequence 1
—, monotone 17
—, — increasing and bounded 42, 88
—, optimal
sequencing 9
sequential decisions in time 6
— — problem 3, 61, 76, 87
— — — with finite alternatives in
  infinite horizon 24
— programming 99–100
shortage, cost 62
—, —, expected storage and 121
—, expected value 63

shortage, probability 62
solution, cyclic 35
—, cyclic native 24
—, "extremal" 130
—, feasible 34
—, uniqueness 44
s, S policy 65–70, 101
stability properties 21–24, 51–53
state of the game 14
— of the system 37, 48
— of the world 1
— probability 39, 58
—, recurrent 55
—, repetition of 9, 10, 15
—, terminating 2, 10
—, transient 55
—, value of a 5
—, variable 78
steady state problems 123–125
stochastic behavior 41
— decision process 55
— matrix 38
— process, stationary 103
stock, receipt of 70
storage 95
— and shortage cost, expected 121
statistics, sufficient 81
Stirling's approximation 81
stock level 64
— —, critical 66
strategy 45

Taylor expansion 114, 126, 128
technical improvements 26
terminal conditions 129, 131, 132
termination probability 123
— of service 113
timing of future orders 63
trade-in value 26
transfer points 14
transient states and all recurrent states
  communicating 55
— — non-communicating recurrent
  states and 55
transition 7
—, finite number of 23
—, probability 37, 38, 45, 56, 58, 76
—, recurrence of 34
—, stationary state, law 54
transportation 11
—, modes of 13
turnpike theorem 96–99

type of investment, point input-
   continuous output 75

uncertainty 5
—, adaptive programming 76–81
uniqueness 17, 35, 47, 71, 90
— of the solution 44
unlimited number of firms 100
utility 2, 3, 41, 96, 101
— function 96, 120
— —, concave 86, 96
— —, identical 86
— —, quadratic 103
— of car ownership 100
— of different periods 85

value, determination equation 80
— expected, of the shortage 63
— function 10, 11, 13, 15, 17, 18, 24, 27,
   41–45, 46, 49, 51

value function, continuity
   of 89
— —, limit of 25
— —, monotone and bounded 46
— iteration 15–19, 46, 65, 93, 100
— — degree of approximation 53
— — no discounting 57
—, limiting 41, 58
— of a state 5
—, present, for the system 109
—, —, of a cycle 122
—, —, of costs 70
variance 101
variation under constraints 132–134
volume of flow 34

warehousing 94
Wiener-Bachelier process 103,
   123
Wilson lot size formula 75

Ökonometrie und Unternehmensforschung
Econometrics and Operations Research

Vol. I   Nichtlineare Programmierung
          Von HANS PAUL KÜNZI und WILHELM KRELLE unter Mitwirkung von Werner
          Oettli. – Mit 18 Abb. XVI, 221 Seiten Gr.-8°. 1962. Gebunden DM 38,—;
          US $ 9.50

Vol. II   Lineare Programmierung und Erweiterungen
          Von GEORGE B. DANTZIG. Ins Deutsche übertragen und bearbeitet von Arno
          Jaeger. – Mit 103 Abb. XVI, 712 Seiten Gr.-8°. 1966. Gebunden DM 68,—;
          US $ 17.00

Vol. III Stochastic Processes
          By M. GIRAULT. – With 35 figures. XII, 126 pages 8vo. 1966. Cloth DM 28,—;
          US $ 7.00

Vol. IV   Methoden der Unternehmensforschung im Versicherungswesen
          Von KARL-H. WOLFF. – Mit 14 Diagrammen. VIII, 266 Seiten Gr.-8°. 1966.
          Gebunden DM 49,—; US $ 12.25

Vol. V   The Theory of Max-Min
          and its Application to Weapons Allocation Problems
          By JOHN M. DANSKIN. – With 6 figures. X, 126 pages 8vo. 1967. Cloth DM 32,—;
          US $ 8.00

Vol. VI   Entscheidungskriterien bei Risiko
          Von Professor Dr. HANS SCHNEEWEISS. – Mit 35 Abbildungen. XII, 214 Seiten
          Gr.-8°. 1967. Gebunden DM 48,—; US $ 12.00

Vol. VII Boolean Methods in Operations Research and Related Areas
          By PETER L. HAMMER (Ivănescu) and SERGIU RUDEANU. With a preface by
          Richard Bellman. – With 25 figures. XVI, 329 pages 8vo. 1968. Cloth DM 46,—;
          US $ 11.50

Vol. VIII Strategy for R & D: Studies in the Microeconomics of Development
          By THOMAS MARSCHAK, THOMAS K. GLENNAN, Jr. and ROBERT SUMMERS. –
          With 44 figures. XIV, 330 pages 8vo. 1967. Cloth DM 56,80; US $ 14.20

Vol. IX   Dynamic Programming of Economic Decisions
          By MARTIN J. BECKMANN. – With 9 figures. XII, 143 pages 8vo. 1968. Cloth
          DM 28,—; US $ 7.00

Vol. X   Input-Output-Analyse
          Von JOCHEN SCHUMANN. – Cloth DM 58,—; US $ 14.50 (in preparation)

Vol. XI   Produktionstheorie
          Von WALDEMAR WITTMANN. – In Vorbereitung